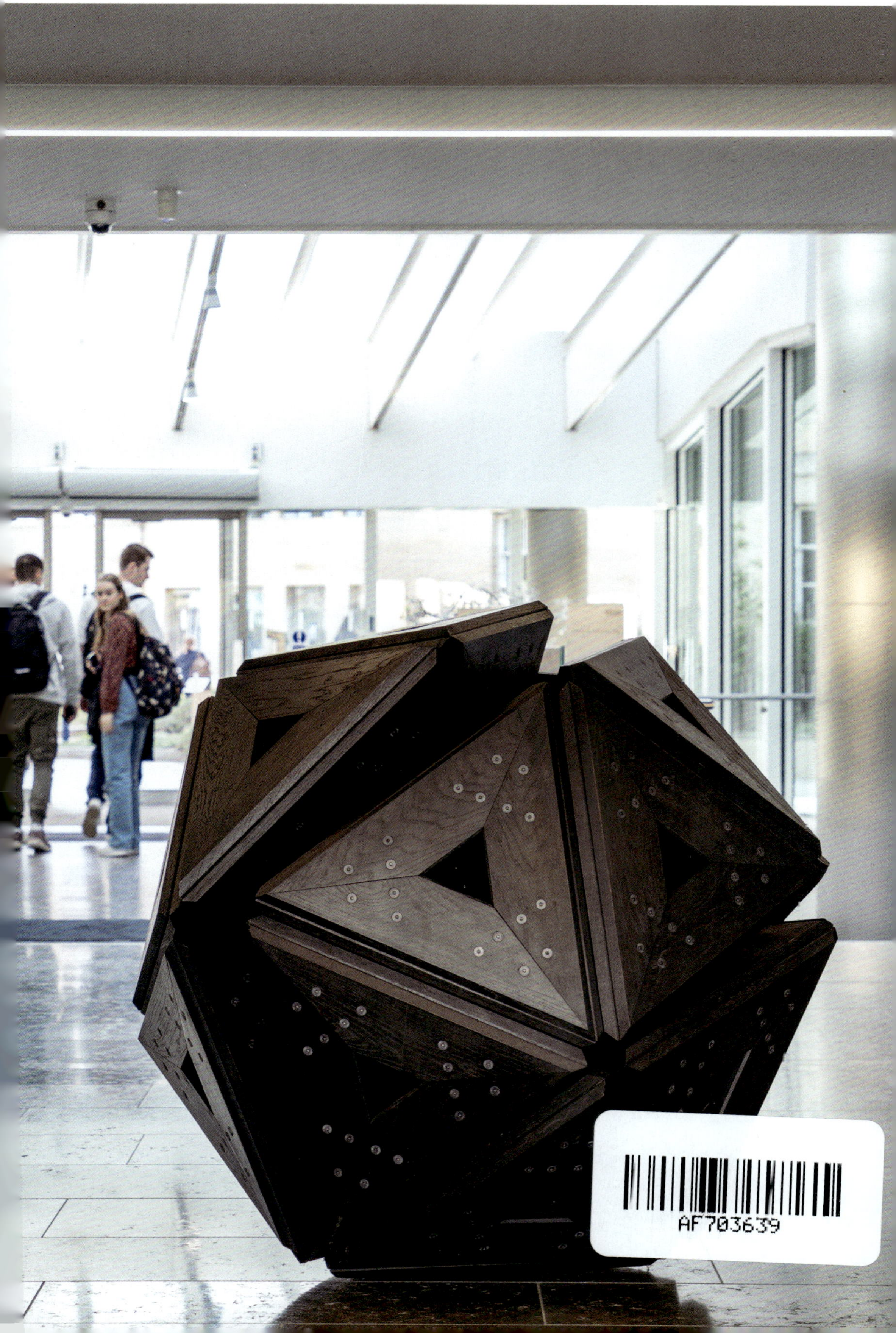

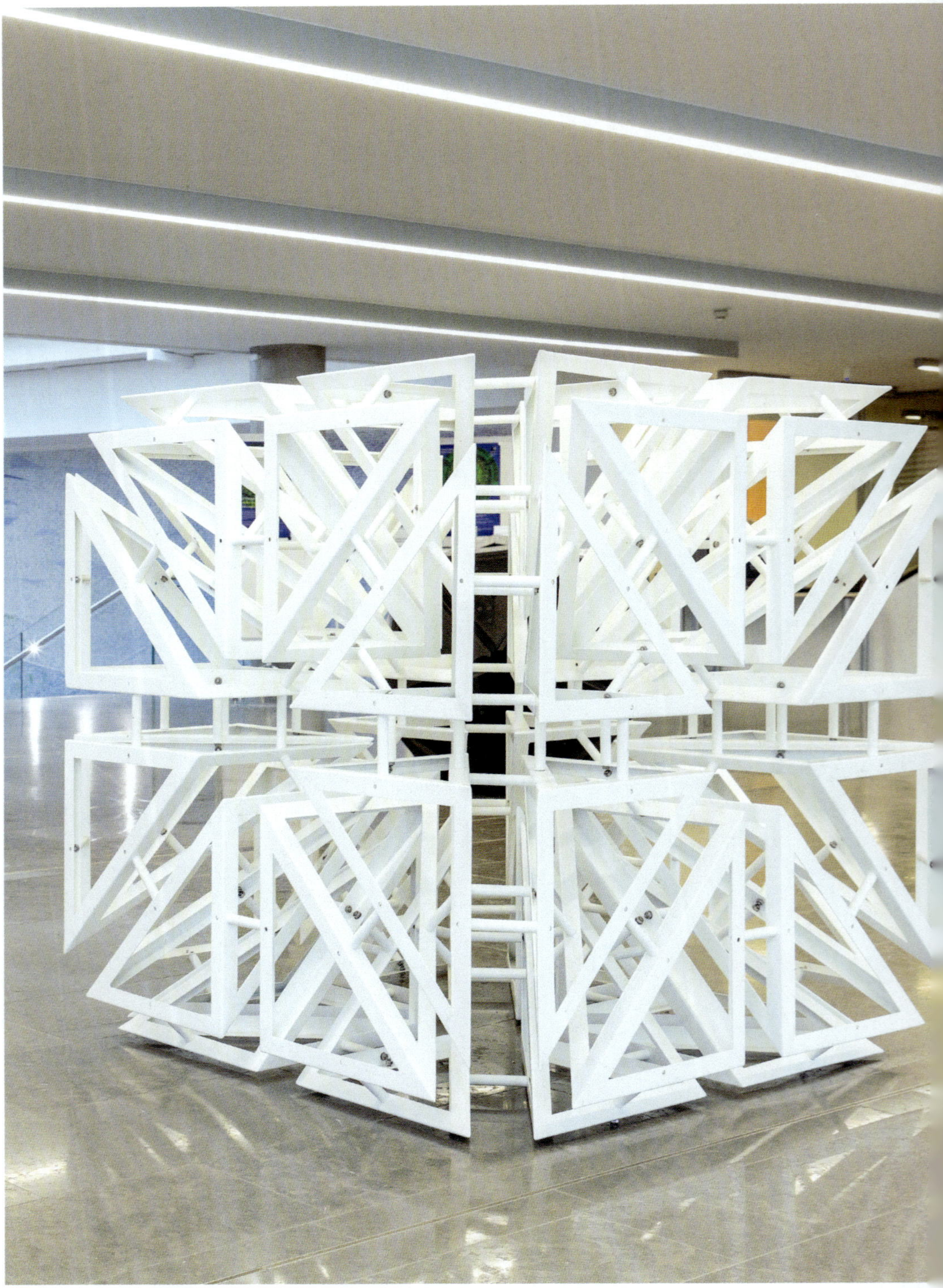

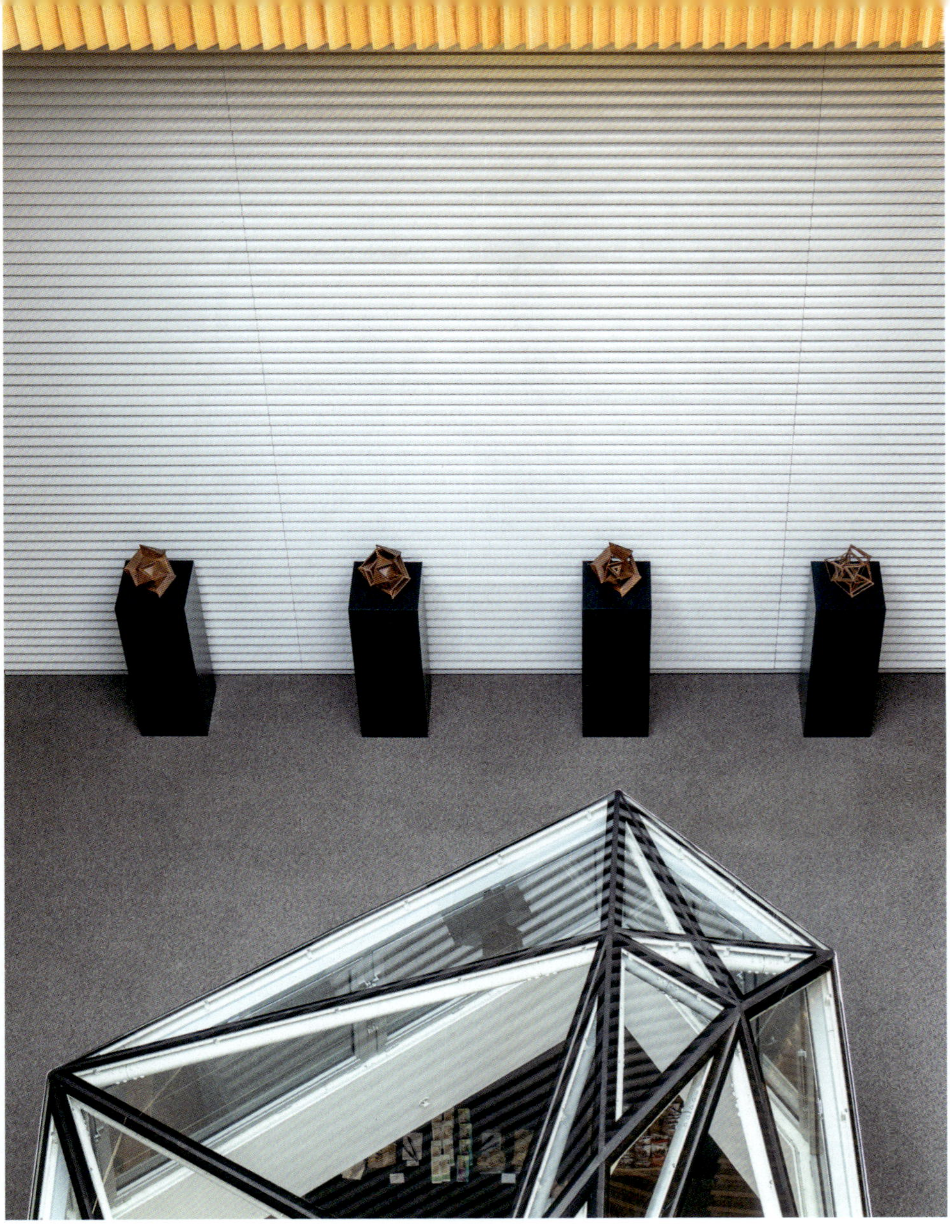

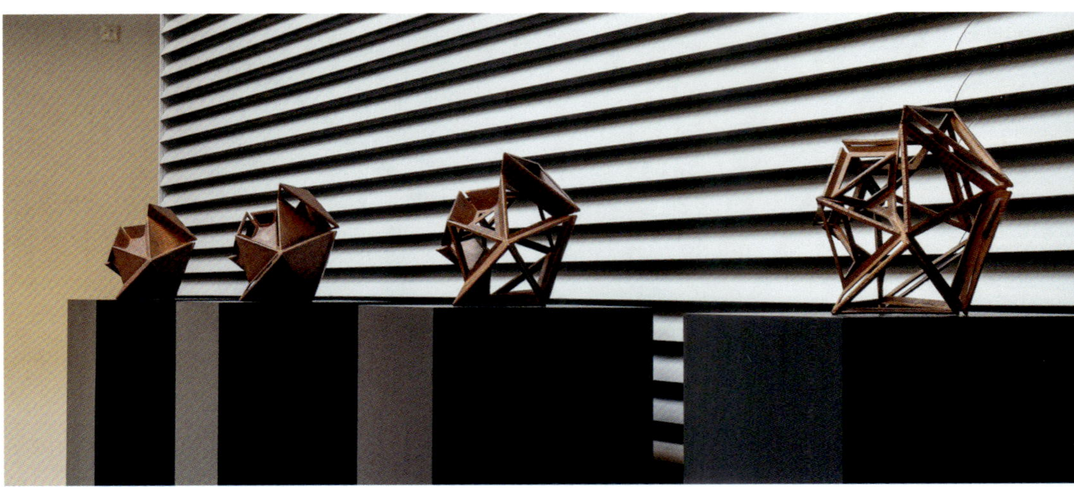

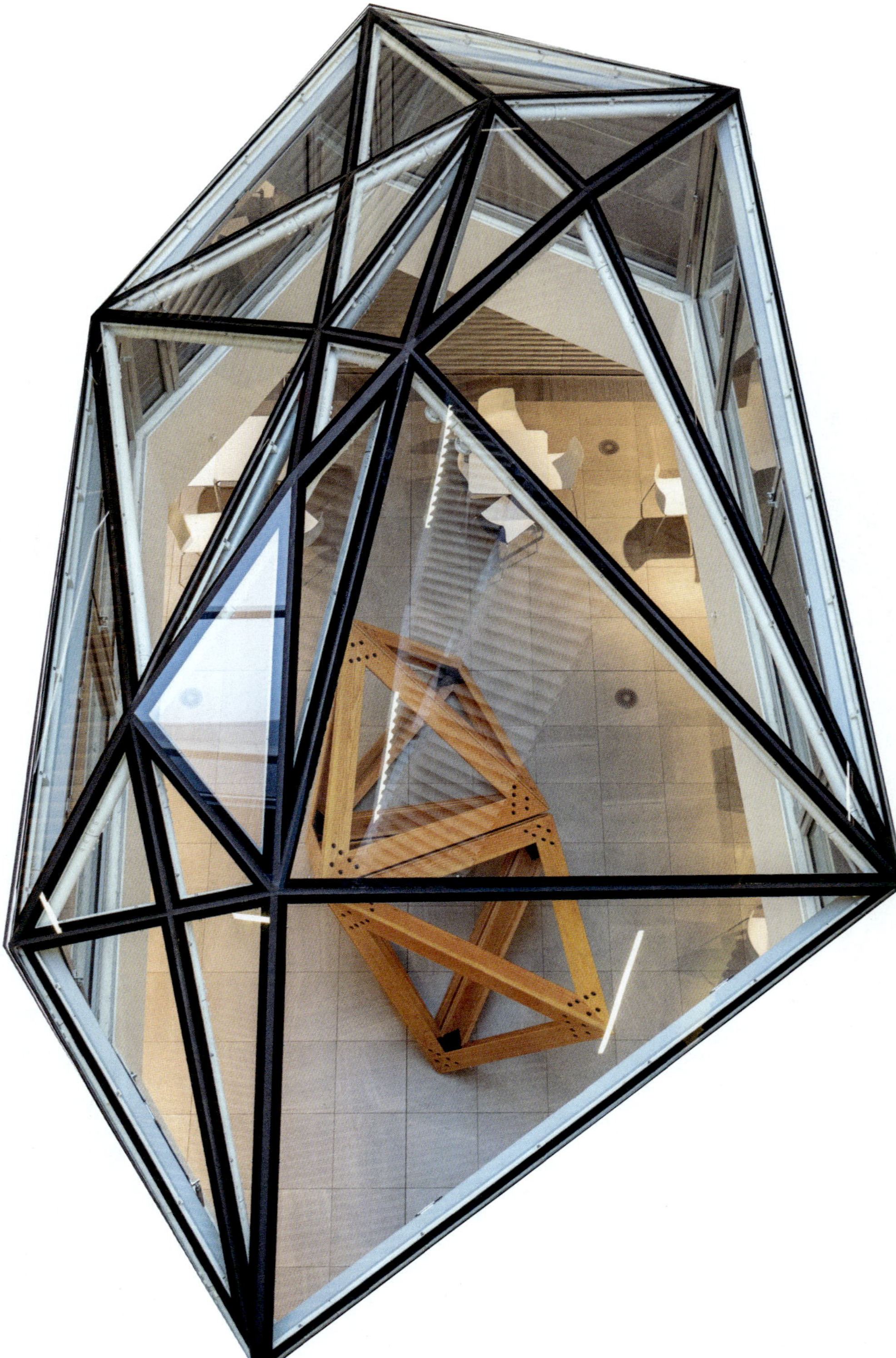

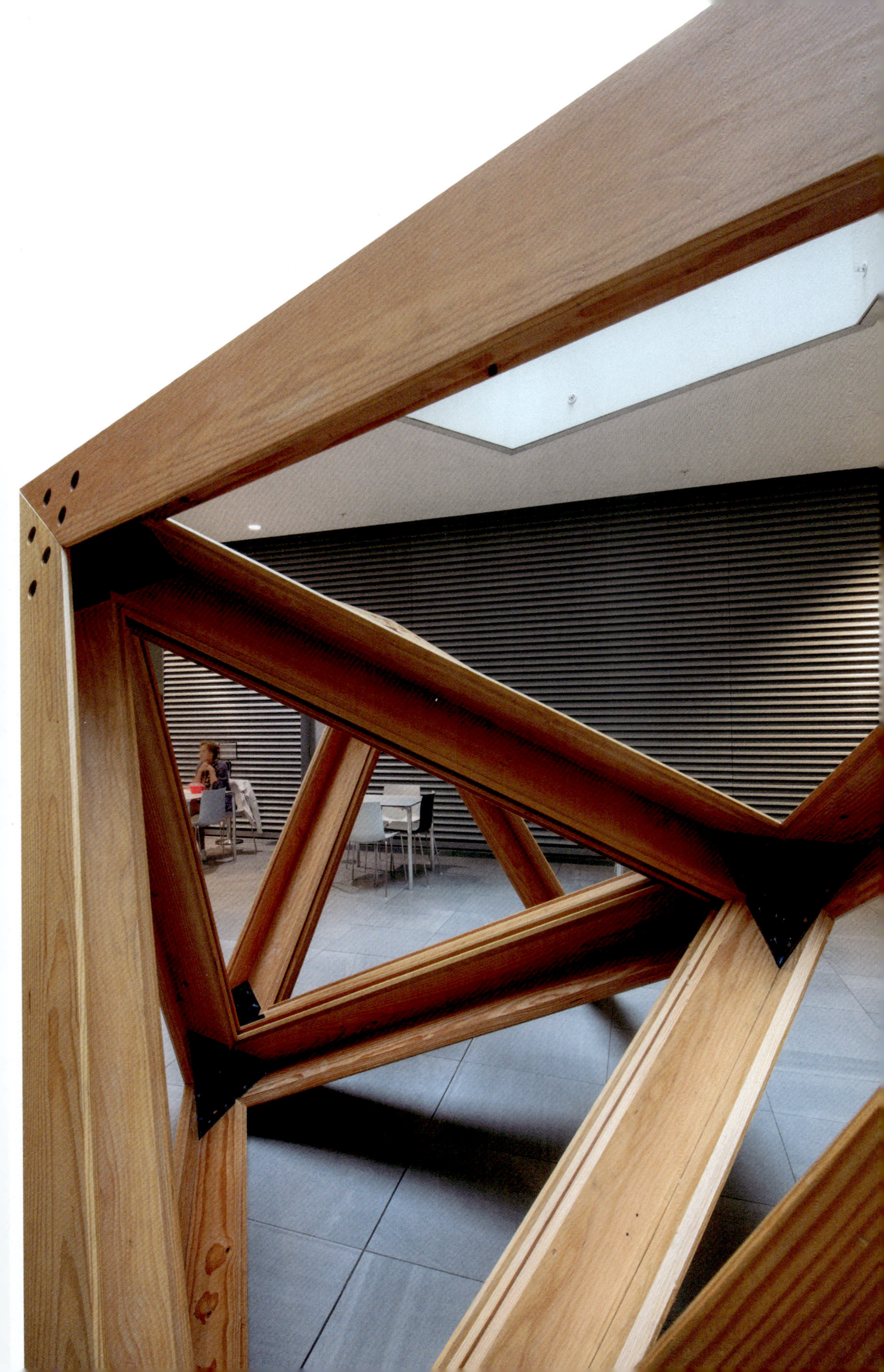

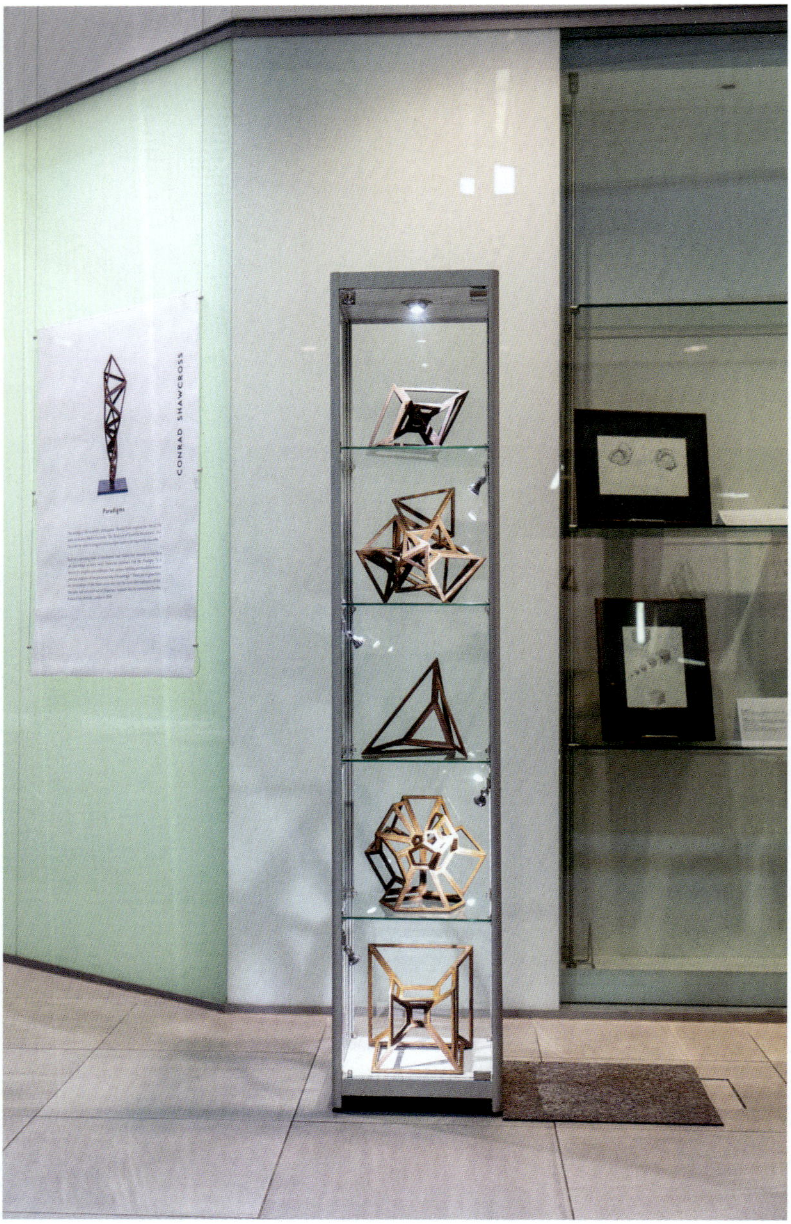
CONRAD SHAWCROSS
Paradigm

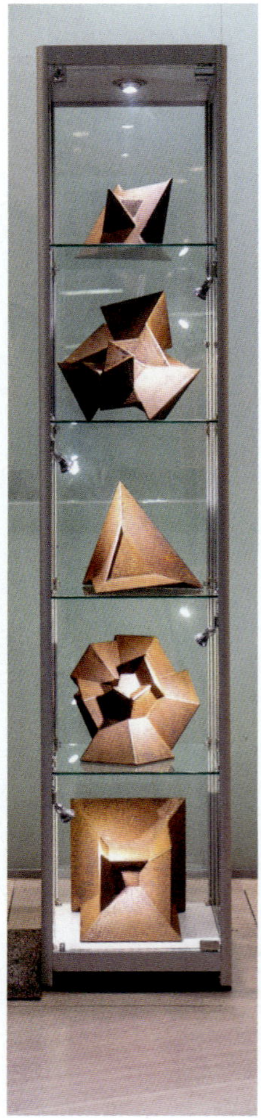

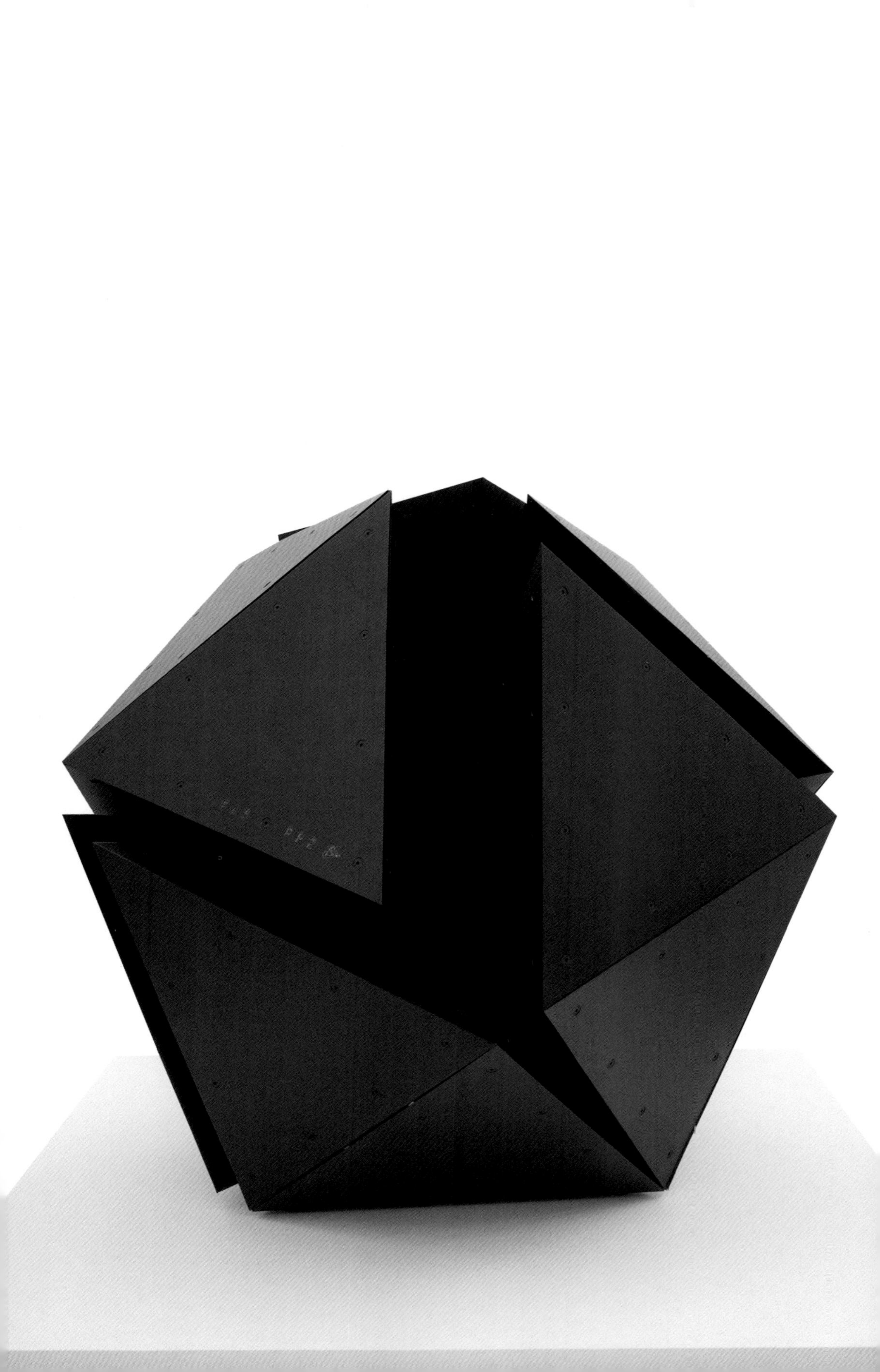

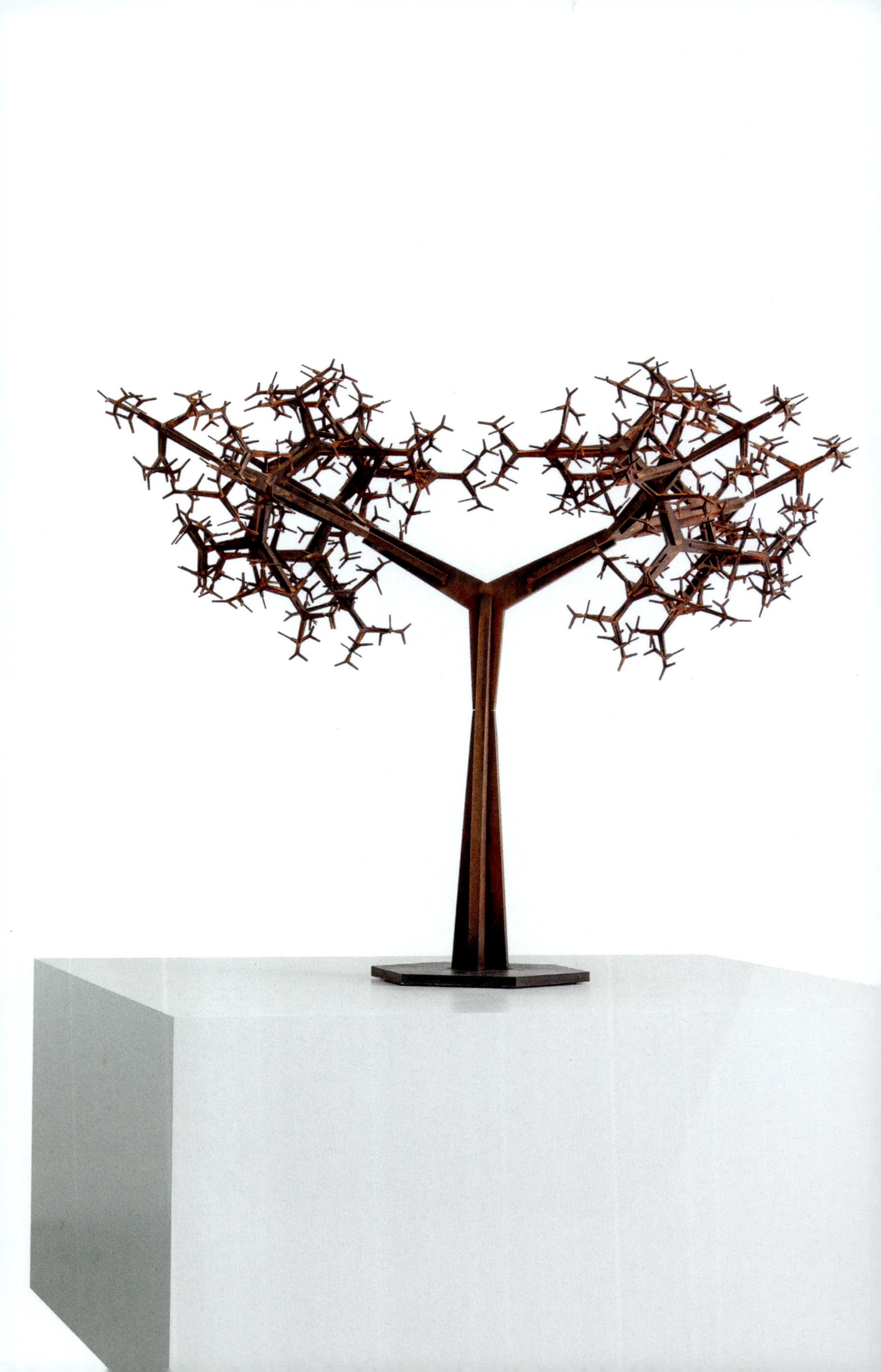

In memory of and in gratitude to my former professor and friend, Brian Catling.

Thank you dog breath for everything.

Foreword

Prof. James Sparks, Head of Department, Oxford Mathematics
Dyrol Lumbard, External Relations Manager, Oxford Mathematics

In early September 2022, our mathematicians returned from their summer holidays to be presented with over 35 works of art spread across three floors of their academic home, the Andrew Wiles Building, works that they soon began to notice had a distinct mathematical structure. How would they react? What would they make of art not only occupying their line of sight each morning as they crossed the threshold, but art which was explicitly incorporating ideas from their areas of expertise? What did they think of the art and, even more pertinently, what did they make of the mathematics? So, we asked them. Actually, in many cases we did not have to ask. They told us, albeit in a polite, mathematician's sort of way.

They had two specific reactions. The first was a sensation plenty of us share in Oxford (and beyond): imposter syndrome. How could they, teachers and students of mathematics, comment on art? They, with the occasional exception, were not qualified. The irony of this was not lost on Conrad himself, given that from the outset he had expressed a certain hesitation about placing mathematically and scientifically motivated art in the working environment of professional mathematicians. But overriding this response was one that in fact answered the first question; and most powerfully, validated all that Conrad wanted the art to do. They liked the maths. No, they loved the maths. And 'they' in this context has few if any exceptions among a department of 650 researchers, postgraduate students, and support staff, and a further 900 undergraduates.

Because the first triumph of this show in all its multi-varied portrayal of order and chaos, certainty and uncertainty, is that it made mathematicians think about their work and so to think about it as art. As many of them said, each day they walked past an artwork they were prompted to think further about their field, whether, for example, in geometry, topology, quantum theory, or even in applied mathematics and its attempts to break down the world into its component parts. And not just to think about it literally, but conceptually and philosophically. And with that triumph came another wider, and most important feature of the exhibition. Just as art inspired mathematicians, so art inspired the wider public to come and see mathematics.

From its conception and initial design, a principal purpose of the Mathematical Institute was to gain the public's confidence about a subject that many shun because of negative personal experience or because they feel they have to be good at it to enjoy it. Mathematics is embedded in the design of our building, from the Penrose tiling out front to the mathematically-shaped crystals inside the building and the Escher-influenced staircases.

The installation of Conrad's work was the culmination of that ambition, mathematical, scientific, and philosophically inspired art, capturing all the ambition and uncertainty of those subjects in a working environment, a gallery of minds alongside a gallery of artworks.

We are proud to be its host and thank the wonderful curating powers of Fatoş Üstek, and, of course, the creative prowess of Conrad himself. Perhaps the final word goes to a first-year student who said, when asked what they thought about the many works that they pass on their way to lectures each morning: 'You mean they are not permanent? It seems as if they were designed for the building'.

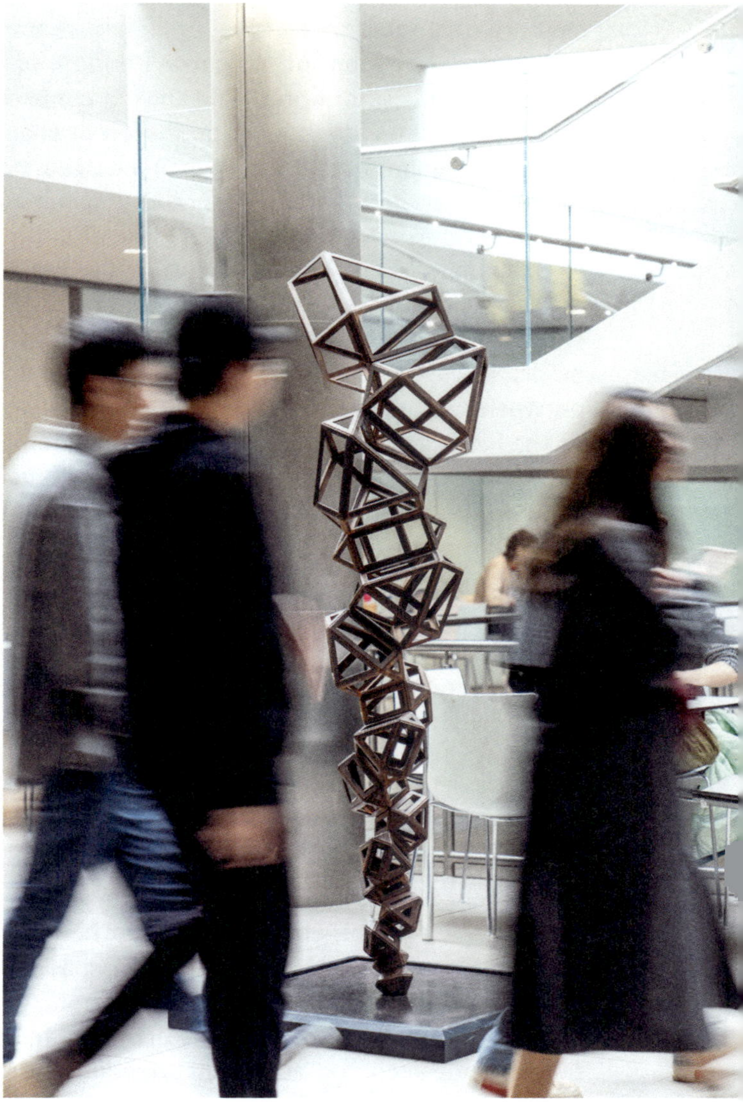

Expanding Boundaries of Exhibition Making and Working Closely with Conrad Shawcross

Fatoş Üstek

'Curating beyond exhibition making' has been a significant motif in my curatorial practice. Working alongside traditional art contexts and activating sites outside the 'white cube' environment has attracted me to projects akin to *Cascading Principles, Expansions within Geometry, Philosophy, and Interference.*

I find great excitement and motivation in exploring the potential of creating art encounters in untraditional art spaces and contexts that are unfamiliar to art objects. These venues could be anything from a pedestrian bridge in central London, to a hidden masonic temple at a five-star hotel. The shift in context is not necessarily bound to the change of location, different scenery, or entry into the public realm; it can occur when a certain habit of curating is accelerated and thereby pushed to its limits (such as *fig-2*: 50 exhibitions in 50 weeks which I talk about in the later pages of this publication).

Context is a strong determinant of exhibition making; it can correspond to a multitude of meanings for the selected artworks on display. One can conceptualise the context as the relationship between factors such as an exhibition space and an artwork, as well as the historical use of a site and the art encounters that take place within it. In certain cases, context can become the content of the exhibition, where the artist provokes its specificity via artworks made in response to its physicality, its features, and its sociocultural circumstances.

Personally, I refer to the context not as the environment in which artworks are placed but as a catalyst for a symbiotic relationship between the artworks and the experiences generated. Hence, when envisioning the exhibition at the Mathematical Institute in Oxford, I took into account the critical charge of the building. I wanted to position the continuous intellectual stimulation and scientific inquiry as a consistent undercurrent of an invisible river that flows through the exhibition. That is to say, I wanted to acknowledge and adhere to all that was in the air: the frequent discussions about topology, toric geometry, abstract mathematics, as well as the daily endeavours of proof-based thinking and evidence backed propositions. Placing a selection of Conrad's works in this environment meant exposing the artworks to an environment that they are not accustomed to and providing the new audiences access points through which they could feel at ease forming their attitudes and formulating their perspectives.

Conrad is a prolific artist working primarily with sculpture. His approach to sculpture follows a nontraditional route of constructing rather than chiseling away from form. His artistic oeuvre is supported by conceptual threads that connect his series of works to one another. He is a polymathic thinker and is inspired by artistic as well as scientific formations of thought. His practice lends itself to diverse contexts, from white cube environments to the public realm. His focus on the tetrahedron – a four surfaced triangular prism – has reverberations both in the arts and in science. Many artists of the 20th century employed a specific form that would inform their wider body of work through use of that form in repetition. In a similar vein, the tetrahedron became Conrad's base unit, the shape that he repeatedly consults in order to articulate and inform his sculptures. In philosophical thought, the tetrahedron is one of the five Platonic solids, which at the time were believed to inform all matter in the cosmos. In science the tetrahedron was once a model for the atom (the indivisible unit of matter) in the period before quantum physics. Conrad's thinking similarly lies at the conjunction of the arts, geometry, and philosophy. It would be a fallacy to think of his sculptures as a bridging of these worlds, however. On the contrary, he furthers his artistic inquiry by producing works that pose more questions than provide answers, without illustrating a specific idea or a certain formation of a scientific thought.

The initial idea for the exhibition was to showcase Conrad's *Paradigm* series in its entirety, and bring together works that span 17 years of artistic production. In its dictionary definition, a paradigm is described as 'a philosophical and theoretical framework of a scientific school or discipline within which theories, laws, and generalisations and the experiments performed in support of them are formulated.' Each paradigm generates a logic for its cause-and-effect relationship. For Conrad, paradigms introduce a framework for perception and experience: hence, in this series, he amplifies different features of his sculptures, composed of stacking tetrahedrons with accelerated growth. For instance, *Paradigm Broad* has sharp proportions due to its growth outwards, and *Paradigm Chamfer* amplifies the welded edges that connect the triangular planes. Each series also exists in structural and solid manifestations, where the volume of these works takes precedence over the relationship between tetrahedrons. The invisible thread that connects all the works in the *Paradigm* series is that they are taken to their furthest points in terms of their construction, with their height defined by their gravitational centres being aligned with the smallest tetrahedron in the stack, the growth stopping before their point of 'collapse'. The *Paradigms*, with their potential 'collapse' coded within, trigger a subtle tension in an environment where scientific thinking is met with set-backs: progress is defined through failures that lead to breakthroughs. Hence, the idea of bringing an artistic position as unique as Conrad's to the heart of an intellectual scientific inquiry into the nature of things, their interrelated states, and the unforeseeable consequences of those interactions, made sense.

The Mathematical Institute is a distinctive building designed by the late architect Rafael Viñoly, who more than a decade ago consulted all the professors and scholars working in the field of mathematics at Oxford University in order to envision the ultimate environment for inspiration and happenstance. For instance, the skylights and the stairs in the office areas and in the public areas are designed in such precise ways that they might spark off an idea in the mind's eye of a student, a scholar, and even visitors. Light moves through the building horizontally, vertically, and even diagonally. Communal areas are carved into the building, where the meeting of bodies (and minds) is supported and encouraged. In such a specific context, the layout could not be generic but is instead responsive, and, if possible, enriching the environment. Introducing Conrad's works to the existing interplay of walls, furniture, and frequent inhabitants necessitated thorough thinking. It might be the building and the meaning that it emanates that fuelled myself and Conrad for such an ambitious undertaking. We decided to add other bodies of work, such as *Fractures*, *Lattice Cubes*, and *Schisms*, which all shared the tetrahedral form in their artistic formulation. The expansion of the exhibition's title took place literally when we included larger scale and kinetic works. For instance, *Off-Cut Axiom* is a work that found its place under the skylight modelled on Pascal's Magic Hexagram. The work itself is an off-cut from an existing piece commissioned by and located at the Ministry of Defence in London, which is placed in alignment with the part of the building that houses mathematicians working in abstract geometry. *Ploison* found its place outside the Institute, in an adjacent square where lunches take place on sunny and mild Oxford days. Furthermore, as with all exhibition making, thematic/formal consistencies need to be broken in order to amplify connecting threads. In this spirit, we introduced works into the exhibition that did not employ the tetrahedron as their building block. Three versions of the sculpture series entitled *Bicameral*, which are made with the speculative bicameral structure of consciousness in mind and using the Japanese joinery technique, were inserted at three strategic points in the Institute: two at the south west entrance and one at the eastern point of the building, in the basement. Alongside the static works, we added kinetic works from the Beacons series, which are each composed of two counter-rotating, circular plates, both perforated by a series of holes that form an interference pattern. These works act as semaphoric signs due to the distinct colouring of their surfaces.

The making of the exhibition had a diegetic process, where we had the freedom to act on the inspiration we gathered from the context of the Mathematical Institute (and unending support from Dyrol Lumbard, External Relationships Manager at the Institute). When we opened in September 2022, we had around 36 works on display. Over the course of a year, we included an outdoor work and four new works. The playfulness of the exhibition took a turn when we started replacing existing pieces with their taller versions and introducing new pieces at unexpected locations, unannounced, at different intervals. The flexibility and openness we received is unique (I do not think it can be matched by a traditional art museum), through which we forged stronger bonds with the Institute and its inhabitants.

To enhance the proliferation of connecting threads between Conrad's artistic practice and the mathematical thinking practiced at the Institute, we organised and moderated (with Conrad) a four-part lecture series that took place throughout 2023. Our intention was to expose ourselves to new forms of thinking and knowledge production which resulted in a series of talks that hosted mathematicians from the Institute, quantum physicists, artists, philosophers, and curators. In order to find a common ground as a starting point to these lectures, we identified four key areas of interrogation to address: principle-based artistic and scientific production, the force of imagination in artistic and scientific process, human yearning for identifying patterns in nature, and lastly the dichotomy between processes of thought generation and their manifestations into artistic and scientific form. Each talk can be watched on the Institute's YouTube channel and they are brought to you in text form in this publication, further complemented by additional essays by participants. Cross-fertilisation has been a key term for the project and the compilation of texts you are holding now is a manifestation of the last 18 months' endeavours.

Cascading Principles, Expansions within Geometry, Philosophy, and Interference is an artistic and an intellectual inquiry into ways of perception and understanding of the world around us through a multitude of sources and ways of knowing. It started off as a seed of an idea that grew roots in the imagination of mathematicians (at first hand) and the wider public who visited the works onsite and attended the lectures. The exhibition became Conrad's largest UK survey exhibition to date, bringing together his artistic oeuvre that spans two decades.

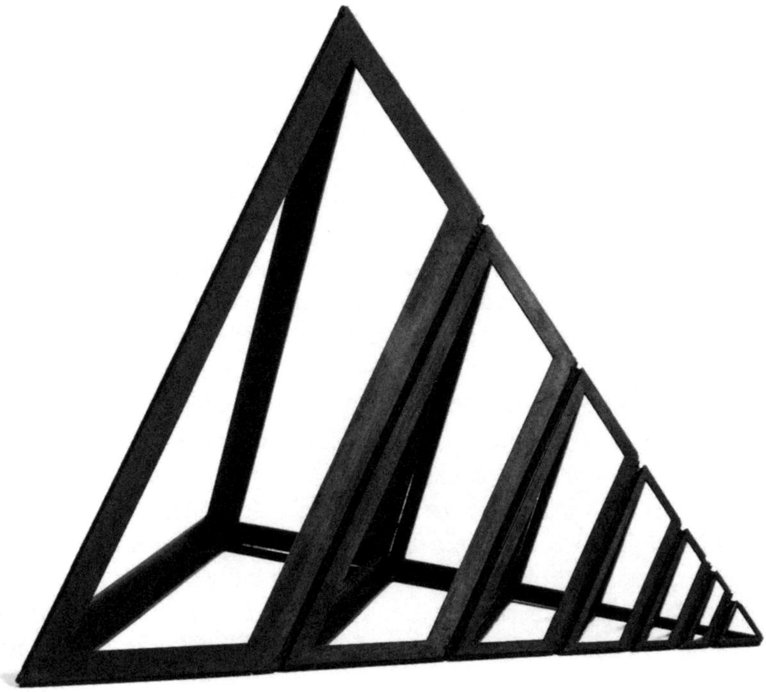

Tetrahedrons

The tetrahedron is geometrically a four-sided form and is conceptually the symbol of an indivisible unit of matter. Unlike other shapes, a tetrahedron cannot tessellate with itself. That is, it cannot cover or form a surface through its repetition - one tetrahedron is unable to fit together with others of its kind. Whilst other shapes can sit alongside one another without creating gaps or overlapping, tetrahedrons cannot resolve in this way.

Chosen as an artistic building block by Shawcross, the tetrahedron behaves as an irrational number, creating sequences which, in theory, extend into infinity without repetition. It is a shape that recurs throughout nature, not least in the water molecule H2O.

Lecture One, 23 February 2023: Cascading Principles

Speakers: Prof. Martin Bridson, Conrad Shawcross, Prof. James Sparks
Moderator: Fatoş Üstek

Fatoş Üstek

I am interested in exploring with our panellists the experience that the exhibition has been proliferating over the last five months. As you know the exhibition is ongoing until almost the end of the year. But before we delve into that let us get to know our panellists. Let us start with you Conrad. Can you introduce yourself to us?

Conrad Shawcross

I am the artist who made the works in the show. I studied at Oxford in the late 1990s at the Ruskin and I was very happy [there]. I had a very positive experience. I had wonderful tutors and I learned a lot. The structure of the University in a way informed the idea for the show in that I/we were spread very thinly, and we still are. The Ruskin is autonomous within the art school, but we are sort of agitators within each college that ask questions, and no one quite knows why we are there and [if] art is academic. It sits a little bit weirdly within all the other subjects. I found that very exciting, to be there, because all my friends were obviously the artists, but I had many friends [who were] writers, [and] people who studied English and History and Science and so I was exposed to all of these subjects.

One of the things I regret is not going to enough lectures beyond those that were dictated by the art school. I did go to a few and they were always revelatory. I think one of the essences of this show is to try and take people into the maths department who would maybe find it scary or foreboding, not appropriate, or intimidating. And just like you, the mathematicians may not want to go to the art school, but it is to try and encourage these migrations.

I think we all probably feel a bit like imposters tonight. Last night I was telling Dyrol that I was nervous about talking in this context. He said: 'Well, that is exactly how James and Martin feel'. I am talking about an area which is not my expertise and vice versa [for the others] but I think that is the whole point, that we are trying to go outside of our comfort zones. Actually, when one is outside one's comfort zone, one can be one's most creative and it is in these moments when real discovery and real progress can be made.

FÜ Thank very much, Conrad. James, could we hear from you and maybe if you can also tell us about what you do?

James Sparks

I am head of the Mathematical Institute. It is a job that I took over at the start of September last year. It has been a very busy six months so far. I was telling Fatoş before we started today [that] it is a hugely enjoyable and rewarding job.

When I get some spare time I also do research. I have been faculty here for the last 15 years. I am a member of the mathematical physics group, so my research I would say lies between theoretical physics and geometry, and I am particularly interested in the overlap of those two subject areas. I think, growing up as a young kid, I was particularly excited by physics and I remember doing a project on the planets when I was about nine or ten years old and that is what particularly excited me. I think I was always very interested in physics despite choosing mathematics as a degree. I thought I wanted to do cosmology. Actually, I know cosmology is also one of your inspirations, the universe and those sorts of ideas, Conrad. Then as a PhD student I got very interested in geometry and so took a slightly different direction at that point. But they very much come together in general relativity. Einstein's theory of gravity is where those two intersect with each other. So that is my research area in a nutshell.

I am also an amateur musician; I have played the organ since the age of 11 or so. I was very much into composing as a young kid but then I would be really interested in the creative process in music and how that was related to maths. That is something I have given talks about, so I am very much interested in mathematics and how it is related to arts and the creative processes.

FÜ Thank you this is so very inspiring, and I cannot wait to collate your responses to the exhibition later on. Martin, we met during the installation and you were very busy organising the Clay Conference. Could you tell us about yourself and your area of expertise?

Martin Bridson

I am the Whitehead Professor of Pure Mathematics here at the Mathematical Institute in Oxford. I am also President of the Clay Mathematics Institute: my main function is to promote, facilitate, enable, and disseminate mathematics at the highest level. It struck me, listening to what other people were saying, that the mission statement of the Clay Mathematics Institute (something I very much like) is to promote the beauty, power, and universality of mathematics, which I think are three themes that go rather well with this idea of cross-fertilisation.

When I am not being President of the Clay Mathematics Institute, I am the Whitehead Professor of Pure Mathematics and my research interests are in geometry, topology, and group theory, which is usually summarised these days as geometric group theory. Some of those words may mean more to you than others. To give you a sense of what I do: I try [to] bridge different fields of mathematics to form fertile interactions; geometry is typically at one end of the bridge. [I] try to recast problems from elsewhere in mathematics or science as geometric problems, so you can think about them with geometric intuition and use the power of geometry and sometimes the other way around. Sometimes you use calculus or algebra to inform you[rself] about geometric objects but more often I am looking to recast algebraic problems, for example, in geometries, so that you can approach them with geometric intuition. In truth that is probably how I approach most things in life, so I am one of the said 'imposters'. I have no particular expertise whatsoever in art but I do have particular reactions to Conrad's work in particular, which is probably informed by the sensibility of wishing to see the world in a geometric way.

FÜ Thank you, that was very beautiful. Perhaps we could talk a little bit about the exhibition; Conrad let us start with you. Why [the] Mathematical Institute, why now?

CS I did a proposal here when the building opened over ten years ago. I was introduced to the building because there was an idea of [commissioning me for] a piece in the atrium. I [proposed] a few quite daring ideas that were completely way beyond the budget at the time. I was proposing a huge Rope Machine that would weave an endless rope and be suspended from the ceiling. While it might have come across as impressive, it was completely unrealistic and so I overplayed my hand, I think. But we remained in touch and something was on the cards before lockdown and then [Fatoş and I] met and realised we had this affinity with mathematics. It sort of snowballed, really, then we met Dyrol (who has been really instrumental), and it really came together.

It started off with a small show of some historic works [which became] a massive survey of all the works I have ever made with tetrahedrons. A sort of tetrahedral Odyssey within the show. There is work to do with interference patterns, all of them again alluding to philosophical ideas. It is that collision of geometry and philosophy that I have a real fascination with. I have this term that I use called 'psychogeometry' which encapsulates the idea that geometry contains meaning, but it may be elusive or everyone who sees it may see something different.

I think the thing that I love about art is that it is not definable, that anyone who looks at a Matisse[1] (like the one outside) will see something different. You cannot dictate that and you never will be able to dictate that. Based on your gender, on your culture, on your religion, on your education, on your love, your trauma, everyone will see something different in that work and it will create different neural pathways in your brain. Everyone will have a different neural response to that geometry or to that piece of art and so seeing these works in this context with all these wonderful augmented minds, these mathematicians, is so exciting to me because I have

1. Henri Émile Benoît Matisse (1869–1954) was a French visual artist known for his use of colour and collage.

a reason why I made those, but it is not as interesting to me as the reasons [for] what people see in it particularly. We have had so many interesting responses from topologists and quantum theorists about what they are seeing in terms of multidimensions. A lot of it I do not understand, but I am so excited by the fact that they see it.

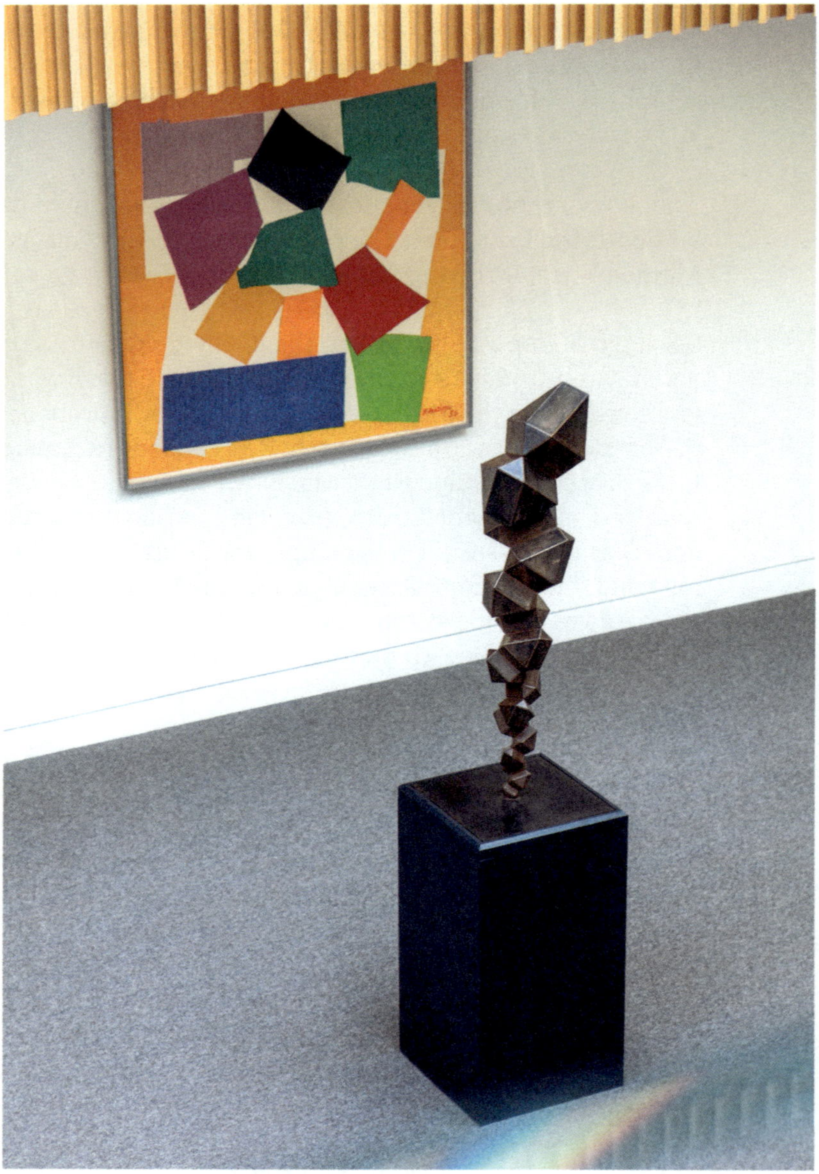

Paradigm Chamfer (Solid), 2015

FÜ Thank you. Martin, because you were talking about your passion for geometry, what does 'psychogeometry' evoke in you? Do you think this is an artistic way of thinking about the multiple possibilities of reading or engaging with something, whereas in mathematics perhaps you do not have that multiplicity of options? Or am I mistaken? If we were to explore this, what would you say is the level of certainty and openness in pure mathematics?

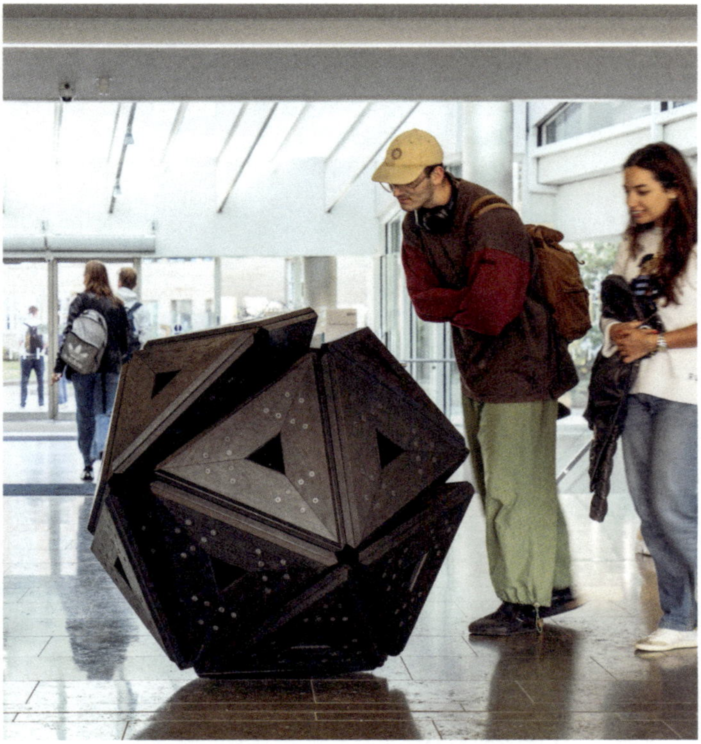

Schism (Black Oak), 2022

MB That is an interesting question. We do look for certainty [in mathematics], we are rather obsessed with certainty, and the proof of certainty. But certainly, what I think most professional mathematicians find most attractive about mathematics is the vast amount of unknowns and uncertainty. The more you try to explore the mathematical universe the more you realise how little we know, how many different interpretations there are of fundamental objects, how many different perspectives there are, and how many different perspectives can be useful in attacking serious fundamental problems. In terms of the art reflecting that, I particularly like the piece in the lobby as you come in, the *Schism*, because I could say on the one hand it is illustrating the fact that tetrahedra do not fit together nicely to tile three-dimensional space, [but] on the other hand it has this real artistic flair about the fact [that] something is almost there... Those cracks are full of potential and that is a very inspiring reflection of my everyday life. Somehow the pieces fit together: you are trying to solve a hard problem, to approach it from all sorts of different perspectives, and in my mind that is really a provocative statement and of course, Conrad has put them together with great artistic flair to bring out a point that I would make in a much drier way. That is the way I think of the jumble of mathematical objects: you have half the story; you are trying to find the other half of the story, to find the proof, to find the certainty, discover the new phenomenon. I find the openness of that work in particular but many of the other works inspiring in that way.

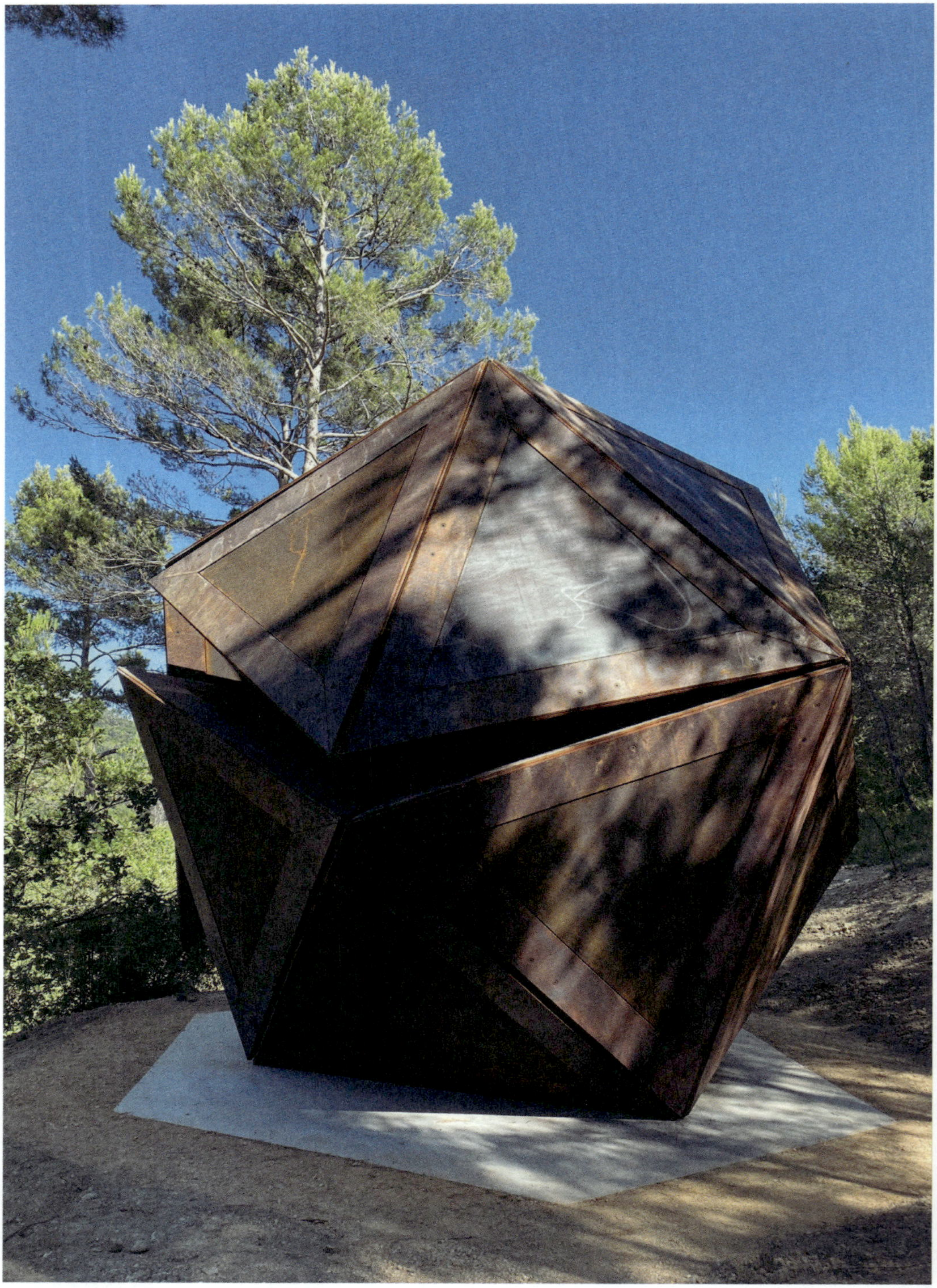

Schism (Château La Coste), 2020

CS It is interesting you brought that up and it is something I discovered ten years ago when I was trying to build a sphere, but I was very bamboozled by the tetrahedron. When I first started using it 20 years ago it looked like this epitome of order because it is made up of four equilateral triangles and it folds up to form this three-sided pyramid. In Greek times it was a symbol of the atom, this indivisible unit, and it encapsulates this sense of simplicity and oneness, but yet when you put them together it becomes chaotic, unruly – it is ungovernable and incredibly surprising. When I discovered [the] contradictory nature of it I put it aside because it was a failure. That perplexed me and I put it aside because I thought this is not working, it is not forming the icosahedron that I was hoping it would, but actually later I realised that is what for me as an artist is so interesting. It creates questions, not answers; it is not a successful model, and for me a good artwork will ask questions. A theologian will see [in this contradictory nature] that God is mischievous, an atheist will say that it proves there is no God, some people say this proves that we are living in a simulation, some people say it is because Europe is screwed. Whatever your beliefs are you see those in it. It is dystopic and it is called *Schism* for a reason, because I like this idea. It is the sort of thing that Descartes would not have showed to the king because he would have been beheaded. Let us forget the art side, would you ever see a model like this in a Maths Institute or in one of those glass cabinets? Is it useful to mathematicians or is the failure just irrelevant?

JS This is also one of my favourite pieces too and for me it represents [that] a lot of our time as mathematicians is [spent on] things that do not work or that you feel should work, just as you described in your process of trying to fit the tetrahedra together. You find that it does not work; you spend so much of your time as a mathematician struggling with things, trying to understand things, understanding really *why* things do not work, and that leads to hopefully a resolution of understanding. I think it is a wonderful metaphor for that, and being stuck, and trying to understand why something does not work or why something does work.

MB I wholeheartedly agree with that and also it provokes the question in what context will it work, what do you have to do to make it work. For example, if you go to hyperbolic geometry and you push those vertices off to infinity you get a very beautiful object, but they do fit together and I think that is the beginning of a lot of mathematical advances: there is something that is just missing there, it is not quite in the right context, so what is missing? How can I fix it? It does not have to be a unique answer, but it has that provocative feel to it.

CS This is the big [version] that I made of that one. The one outside is made of wood, but this one is in the south of France in a place called Château La Coste and it is a 6.5 metre one where there is one tetrahedron removed so you can actually enter into the crack, there is a space inside. I did this during lockdown, all these planks of steel [arrived] and there was no one to install it. I drove out, got a licence to do it, and trained these Argentinian pizza chefs to help me put it together. I have become the colour of the steel. It was a lot of fun and we had to use a lot of improvisation.

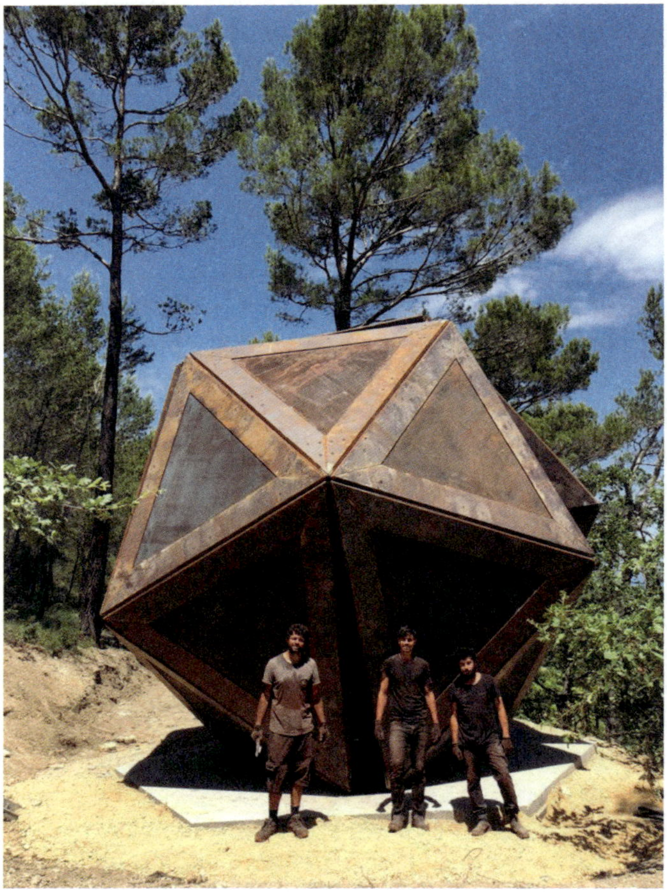

Schism (Château La Coste), 2020

I was told (and this is a question for both of you) [by] a mathematician who I will not name, that there is a certain condition in the universe (in terms of the curvature of spacetime) [which means] that these cracks would not be there. Is that preposterous or is that a possibility?

MB That is the hyperbolic thing I am talking about.

CS Yes, so there is a condition?

JS In a negatively curved universe (which we may or may not live in) that crack could close up if on a sufficiently large scale.

FÜ I think that is very beautiful. It provokes many ideas regarding what we perceive as reality and the context in which that perception is formed.

MB [I want to pose] an artistic question: [When putting 20 tetrahedra together], there have to be gaps but one can fiddle with them to put the gaps in all sorts of different places. How long did you stare at it and fiddle with it, and are you always content that you have come to the optimal answer?

CS I do not know how many ways you can put [20 tetrahedra] together, but every time you put it together it forms a different crack (there are nicer ones than others). This is all about the human experience, so the voids within it needed to be approached at around 1.4–1.5 metres (at eye height) so you can navigate that. I did not want to have any spikes coming off at a dangerous height, which is between knee height and eye height. There were all these things you have to take into account when you are making something at scale. But scale is something that is important with the way these works scale up.

FÜ Conrad, perhaps it is a good time for you to talk about how tetrahedrons as a form, a unit, informed the wider oeuvre of your practice?

CS I have been on this journey with the tetrahedrons. It has been this most amazing voyage, because it keeps on giving and I keep discovering things about it that baffle and confuse me. This is such a rich vein and it is a really good example of how important teaching is.

When I was first taught about the platonic solids at primary school, I was told there were five of them [with a certain] number of sides, and that was it. Actually, I am incredibly flattered [that] my maths teacher from high school is here, Mr. Jones Parry. I think he was the one who redefined them to me. Actually, there [are] only five of them because [of] the only ways [in which] you can divide the surface of the sphere up equally. For me this completely shifted what they meant because essentially you can inflate a tetrahedron to form a sphere. You can blow it up, and you can do the same with the cube (which seems counterintuitive) and you can do the same with the dodecahedron, and the octahedron, and the icosahedron. They are all these radiant geometries; I started to play around with that, pushing and pulling the forms in and out. That was a really interesting thing about this, but I loved the tetrahedron also because it was the symbol of the basic unit of matter. It was like a brick, the sort of quantum brick that I could use, like Sol LeWitt[2] and [other] rule-based artists. I loved these [artists who worked with] units and I wanted to find my own unit, so I took the tetrahedron, which was the most complicated shape to choose as your brick. It is very difficult, no one builds with tetrahedrons for a reason, but it has been amazing and it has been

2. Sol LeWitt (1928–2007) was an American artist linked to the conceptual art movement and minimalism.

a revelation and been bamboozling and frustrating but it has been extraordinary.

MB It is a bit of a tangent but I think the tetrahedron is a fascinating [form]; it is somehow the most natural platonic solid.[3] It is the one that most naturally exists in every dimension. The other [solids] do not generalise to all dimensions – the cube does, but not in such a natural way as the tetrahedron, which exists equally well in other curved spaces. In fact, if you take a so-called manifold in any dimension (a space that locally looks like Euclidean space), one of the most famous theorems of Whitehead (the name attached to my chair) is that any compact model of space in any dimension can be divided up into tetrahedra (the equivalent of tetrahedra end simplices) in that dimension (but you have to bend them in a curved way). The fact is that any reasonable space you are going to come across in mathematics can be so-called 'triangulated', can be divided up into the higher dimensional equivalent of tetrahedra. Crucially, you have to bend them, which is the fascinating thing where you move from geometry to topology. It is a topological fact that you can do this and then the geometry comes from what shapes you have, to make these tetrahedra.

CS When you say bend, do you mean they have curved sides or they are tetroidal?

MB [I mean] that everything about them is curved so you can do it in three dimensions. Start with four points, bend the edges between them and then you are allowed to curve the faces as well, and you would have to do that in every dimension; you have to allow that curvature.

FÜ James, how does this sound in your area of exploration, the idea of curvature and would you like to add something?

JS All sorts of things sprung to mind when I first saw these pieces, Conrad. Following on from Martin's comment, simplicity or complexes in topology. If you are a topology student one of the first things that you learn to do is divide up your space into simplices, which are the higher dimensional versions of tetrahedra, and tetrahedrons, the three-dimensional simplex.

I also work in an area called Toric Geometry[4] where exactly these sorts of tetrahedron polytopes also enter into. I happened to be working on that at the time when you were moving the pieces in last August so that was the first thing that came to mind. For me

3. Platonic solids are any of the five geometric solids whose faces are all identical, regular polygons meeting at the same three-dimensional angles.

4. Toric geometry lies in the overlap of algebraic geometry and polyhedral geometry.

these represent seven dimensional spaces that I am interested in and the three-dimensional space that you are seeing, the polytope, is some particular projection of that, where you are seeing an image of that in three dimensions, but the polytopes capture that seven-dimensional space.

I think, to answer the broader question that you are asking about this particular topic, one thing we have not touched on very much so far is the aesthetic nature of mathematics and how important that is. There is an inscription outside the Maths Institute when you walk in over the Penrose tiling which says: 'To freedom and the pursuit of beauty in mathematics' and so beauty and aesthetics is very important, particularly in fundamental mathematics. I would say it is one of the drivers of that. If you think of what a mathematician do[es] you might say well it is something to do with equations but we do not just generate true equations, that is not what mathematicians studying the foundations of mathematics are doing. They are interested in relationships between different areas, particularly striking, unexpected relationships between different areas. I think the sort of things that mathematicians are naturally drawn to and find beautiful and elegant is when you start with a very simple, elegant, and natural starting point and you explore its consequences, particularly when there are unexpected consequences.

FÜ Could you perhaps give an example?

JS There is a very good example of this related to Group Theory. The study of symmetry and maths is called Group Theory and you can make an abstract definition of what a Group is, so you can think of it as like rotations of an object in space. That would be an example of a Group. It took around a century for mathematicians to classify the finite simple groups. Finite means they have a finite number of elements in them. Mathematicians have a habit of taking English words and then making very specific definitions [of them]. 'Simple' Groups has a very specific definition. It does not mean that they are somehow simple. The surprising thing is, despite the fact that the definition group is very simple (so you could write it on the board in a couple of lines), there are hugely complicated examples of these things that took a whole century to figure out. There is this thing called the Monster Group which is very famous (it is absolutely enormous); it has got something like 1,053 elements in it (that is more than there are atoms on the planet). It is incredibly complicated how this Group fits together, and its structure, and you have no idea that that structure was hidden in the original starting point. I think that is the sort of thing that mathematicians find beautiful and appealing: when you take a natural, simple starting point like your tetrahedron and then explore the consequences of that and the process of discovery, of trying to understand the consequences. I think musicians go through a similar experience when they are writing music too, of taking a simple starting point and exploring its consequences.

CS Yeah, I think technically or realistically I am not a sculptor, I do not take stuff away. It is an additive process and so I take these bricks or these shapes and quite quickly, from the repetition of that shape, you end up with very complex situations. It is that kind of rapid proliferation into complexity through these quite simple progressions, but they are not sculptures, they are progressive additions or ascending stacks. Sculpture is an inaccurate term for a three-dimensional practice, but in essence it is an additive process.

FÜ I wanted to also bring this conversation [around to] the building we are in and the fact that we have been imagining how the works would be part of the everyday social texture of students, lecturers, conference panellists, mathematicians, researchers, and Oxford residents. I really feel romantic about the fact that in the canteen people can have lunch with the sculptures. I think that is really not common in the art world, we do not necessarily dine with sculptures.

MB I think you make a very good point about the conversation between the building and the artwork. Part of my enjoyment in it, but also I think why it works so fantastically well, is that this building is special. It is not a building that mathematicians just inhabit. This building was built for mathematics and I think that really shows in the architect. The architect Rafael Viñoly really had a proper conversation with us about not just what nice working conditions [would be] for mathematicians but about the nature of [a] building that facilitates mathematical interaction, that welcomes people in, that gives people a sense of the importance and the excitement of mathematics, and makes a statement to the general public: look, mathematics is not this dusty, dry, dead business, it is this exciting thing that is happening here. Part of that is the light and the beauty of the building, but also it is structured for interactions [and] has mathematics embedded in it.

Some of you may not know these two beautiful crystal [skylights which let] natural light into the basement also have a real mathematical meaning. One of them has an ancient theorem of geometry, Pascal's theorem,[5] encoded into it, the other is an eigenfunction of the Laplacian (that is to do with the problem of 'can you hear the shape of a drum'). I think having Conrad's work here just extends that feeling, that it is art for mathematics, art

5. The north crystal embodies two very different pieces of mathematics. One subset of the structural members trace out what is known as the 'Utilities Graph'; a familiar problem in which you are asked to join three houses to each of gas, water and electricity without the lines crossing. It also includes the theorem from geometry known as 'Pascal's Mystic Hexagram' which says that if you inscribe a hexagram in a conic section then the points of intersections of the sides lie on a straight line.

of mathematics, art in relation to mathematics, and it seems to be a perfect venue to have that conversation with the building and the people in it.

FÜ Thank you Martin. James, would you like to add something?

JS Thanks, Martin. I fully agree, it is almost like this building was purpose-built for this particular exhibition, but it is really a very unique exhibition. We have art exhibits in places where students are working. We are in L1 at the moment (this is where our first-year undergraduates come every day for their lectures) – there is the cafeteria outside – but then there are artworks there. It is also where we have our mathematics seminars; people come out and your *Axiom* is right outside, where students are sitting and working on mathematics problems. It is really a very unique thing and it is hugely exciting to have it here.

FÜ Thank you for your wonderful elaborations. Conrad, do you want to perhaps walk us through the show and share some of the contexts that your work has been expanding on?

CS I thought it would be good to run through some of the slides, some of the journey through how the works have progressed and arrived.

This is a really old piece that I think I did in 2005 and it is the most successful failure that I have ever made. I put it in because I do not regard this as a finished artwork, but I learned so much from it. This was where I had this quite hubristic thing where I made 800 tetrahedrons. I arrived with a truck in this garden and I had two weeks to put two sculptures together. I started to try and work out the rules and logics of the system, then was completely bamboozled by these anarchic forms, and I just could not find any [consistent way of putting them together]. It just became these fiery tendrils and it basically was controlling me rather than me controlling it. I did what I could and made this form. It was also structurally very precarious, but this is where I discovered the tetrahelix, it was not something that I found by research. I later realised that [Buckminster Fuller][6] had coined this term, the tetrahelix, and so in the middle of this you can see there are three tetrahelixes that come off the bottom. Then I built clusters that were all identical that I bolted on, so there are attempts at modularity and rulemaking, but the three tetrahelixes are at the core of it. I was also beginning to feel frustrated that the tetrahedrons in the centre were the same scale as the ones on the outside, so I wanted to convey a sense of entropy or change or there being older tetrahedrons in younger tetrahedrons. There is this sense of time or growth or expansion so this one where there was no

6. Buckminster Fuller (1895–1983) was an American architect and 20th century inventor.

difference between the middle and the outside was beginning to gestate in my head, crystallise in my head that that was not correct.

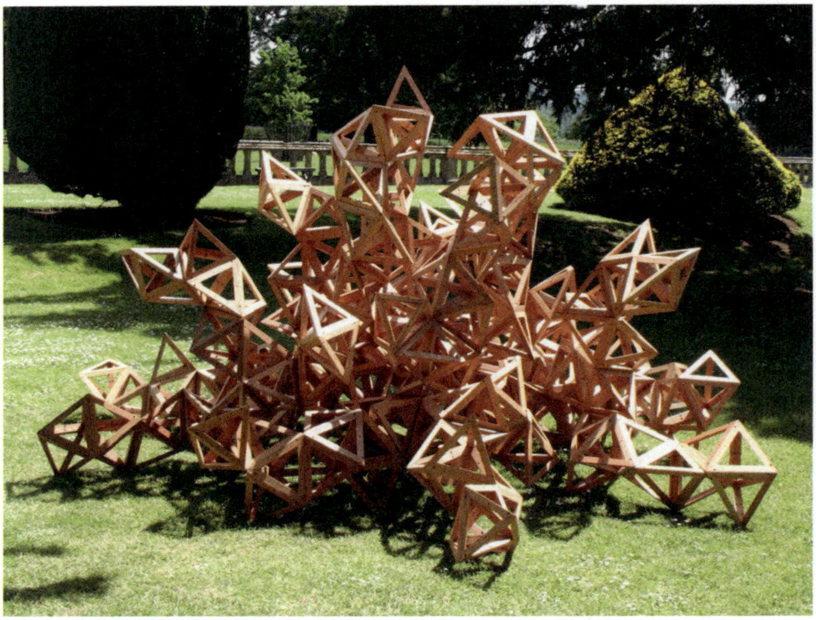

Space Tessellation (Triple Helix), 2005

A couple of years later [2007] I got this commission [to install a sculpture at] the Ministry of Justice [in London]. This was a little bit of a Trojan horse because I told the Minister of Justice this was all about stability and strength and certainty but, actually, it was all before the financial crash and everything was feeling very precarious. I was really excited about the fact that this felt very precarious and I was like if we add another tetrahedron to the top the whole thing will collapse. So, there is this wobbly tower in the middle of the Ministry of Justice which still makes me smile because it is the opposite of what they should really want to convey there, but it is still there. The pieces in [this exhibition, located in the basement, on the right side of] the café, are the off-cuts from [the sculpture situated at the Ministry of Justice]. We had all these planks left and I made this piece called *Off Cut* [*Axiom*] which is on its side.

What is exciting about when you scale the tetrahedron up [is that] it means something so different. When you see these small things on a table and you scale it up and it becomes human sized, architectural, or immersive. The implications of it to one's body are so different and the meaning is so different; scale affects meaning. With that piece there was again this sort of frustration that the top was the same size as the bottom. With Albertian principles a facade will be slightly bigger at the top to make it look square to the eye. What I realised was that I need to start messing with the geometry in order for them to look right, so then I started to expand them and create tetroidal forms which grew at different rates. Computer aided design was really

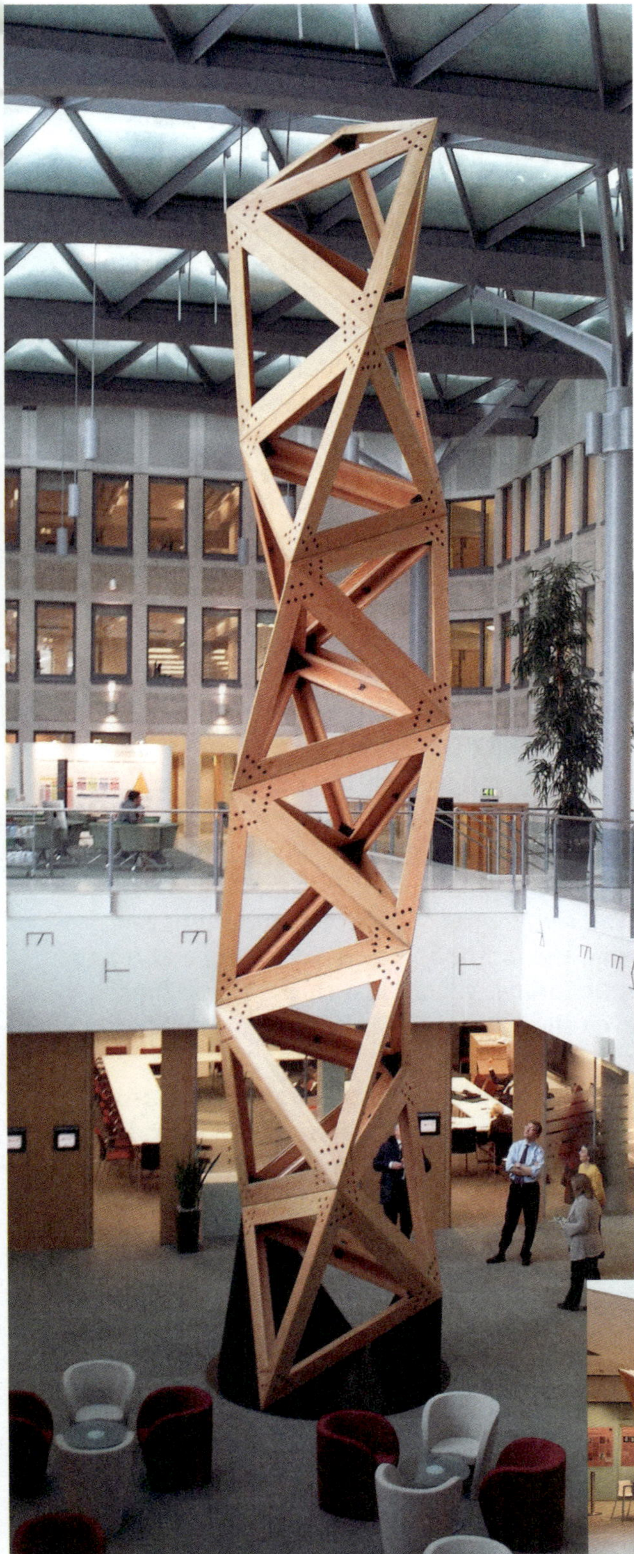

Axiom (Tower), 2009

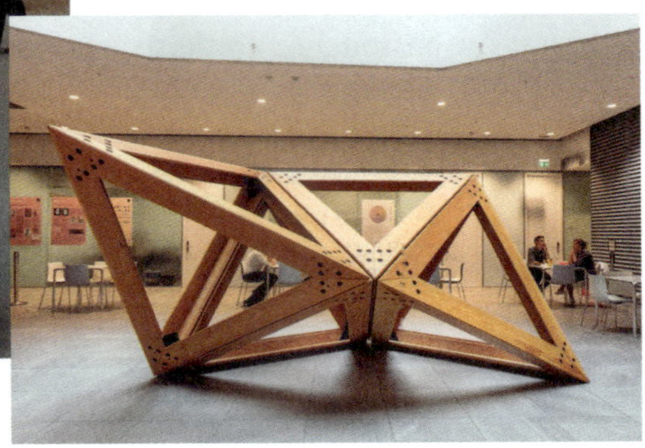

Off Cut Axiom, 2009

instrumental in this. I am not sure these sorts of forms would be possible without computer aided design; it is so complex, the geometries of it all. I am sure it is possible but it is just that it is so irrational and the interesting thing about the tetrahelix is that it never repeats, so it never comes back to the same point. There is all this quite irrational stuff within these things which happens so quickly when there is so much order, when they are on their own, and so much chaos when they are together; it is so quick to occur. Then all these explorations led to different growth rates. There is a Broad one in the background that is accelerating at about 20% per tetrahedron. This is a Chamfer where the welded edges have been accentuated massively to create large triangles on the tips. But essentially it has still got the tetrahedral dogma. The interesting thing with these [*Paradigms*] is that in terms of an ascending form where you imagine time moving up, there is a sort of weird contradiction in that the [smallest] is the eldest and the largest is the youngest so you have got this strange counter intuitiveness which I think makes it more exciting sculpturally.

We have basically put all my artist proofs in the show, 16 of these *Paradigms* dotted around in different locations. Fatoş and I chose all the locations and you are both right they look wonderful here, the visual context or palette is great. The conceptual context of the people who see it is the thing that excites me and all the conversations or the ideas that may come from them. [It] feels like they are in a really good place.

FÜ We are going to explore the *Paradigms*.[7] Earlier in your journey with tetrahedrons you were using the methodology of repeating and trying to map a surface with tetrahedrons. Whereas with the *Paradigms* series in the exhibition, even though there is repetition of the form, as you mentioned, the scale is different.

During the install, some mathematician scholars would approach us to share the ratio of growth that they calculated on the spot. It was rather an unsettling experience I must say. It is quite interesting to install an exhibition in this context because everybody immediately tends to calculate, and measure while engaging with the sculptures.

I would like to contemplate the 'danger' that you inherently introduced to these sculptures. Because, as you say, one more tetrahedron would make it collapse.

CS Absolutely, and all of those pieces came out of research because I was given this commission at the Francis Crick Institute (again a scientific context), opposite the Eurostar in King's Cross [in London]. Sir Paul Nurse was very instrumental, and he was very excited about this idea [of paradigm collapse]. I proposed the word *Paradigm*, [bearing in mind its etymological meaning and its conceptualisation

7. Paradigms are a series of sculptures made by Conrad Shawcross.

by] Thomas Kuhn.[8] The [Institute] is trying to solve cancer and [other diseases] – it is a wonderful laboratory of scientists. The premise of this piece [is that] it is a stack of tetrahedrons bolted together. The piles go down 30 metres, the sculpture is 14 metres tall, and if you added another tetrahedron to the top (which would be the size of a double decker bus) it would fall over. I wanted [to imply that] progress leads to collapse. The collapse is important. Particularly within science you have to chisel away at ideas of certainty and it is not arrogant for the young students to question the things that are believed to be true. The only way that science can progress is by interrogating, cross-examining, and questioning certainty. If this building was to succeed, the sculpture collapses and progress is also intertwined with collapse.

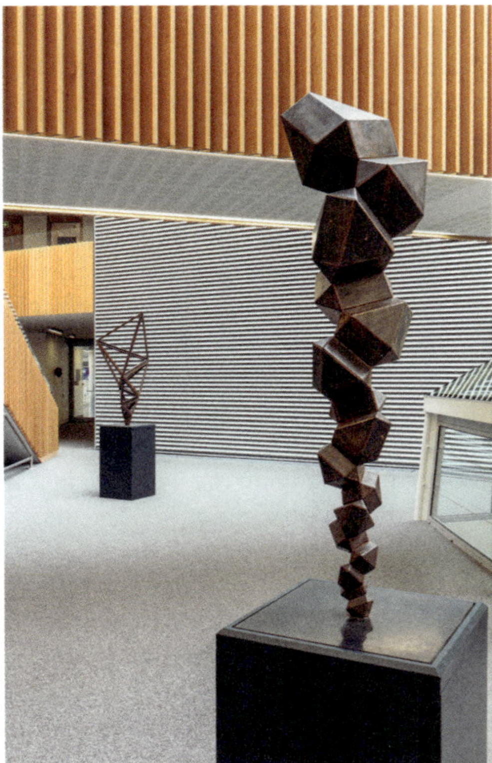

Paradigm Broad (Structural), 2016 and *Paradigm Chamfer (Solid)*, 2015

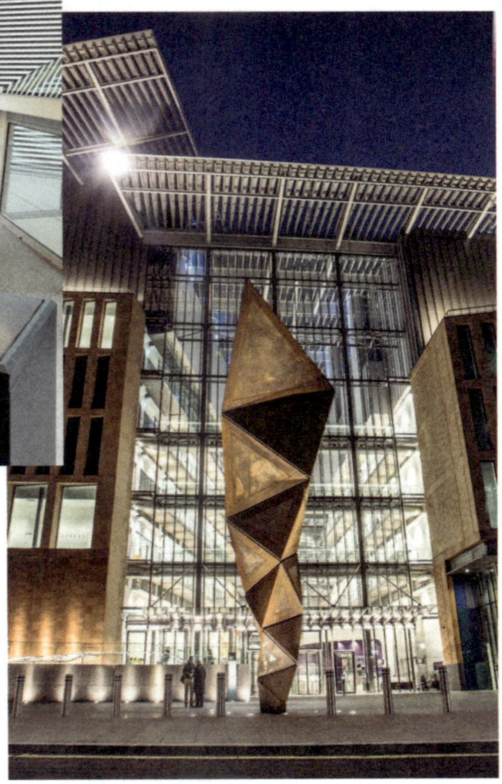

Paradigm, 2016

8. Thomas Kuhn (1922–1996) was an American historian and philosopher of science who introduced the term 'paradigm shift'.

JS With my head of department hat on I was a little bit worried about the [*Off Cut*] *Axiom* and whether that was likely to topple over, especially if it is nudged and squashes an undergraduate.

CS It is very stable. I have to say, these works are all completely underwritten by my structural engineer who is extremely good, so they are all very safe. But I do push him hard to make things feel like they are working very hard and the materials are honest, there is no scaffold down the middle, it is exactly what it is. I would never do something that has a prop. It is doing what it looks like it is doing and you cannot really fake that, in my mind.

FÜ Thank you Conrad. What about in mathematics? Is there any 'danger' we should be aware of?

JS There is no inherent danger. Usually doing mathematics is fairly safe, you work with a pen and paper, but what I would say is that it is important to be ambitious in doing mathematics and I would especially give that advice to young students who are embarking on a career in maths.

I think it is quite easy to become stale in your subject area where you do not venture outside of that. I think pushing yourself as a mathematician is an important part of that process, so I certainly resonate with that.

MB Sadly, and like many people, I am not good at knowing when to give up, knowing when I have not solved [a] problem in ten years and should probably stop. Maybe I am banging my head against a brick wall that really does not exist, maybe it is not just one more idea, maybe there actually is a natural limit to my method that is not going to solve this problem. There is some intrinsic difficulty that just cannot be overcome, the problem cannot be solved in this way, the final tetrahedron will [need to topple the sculpture] over before you get where you want.

On a different topic, Conrad was saying to me earlier how ultimately the piece, like a piece of poetry, has to speak for itself. I must say, hearing from the artist of the idea of the conception [of an artwork] adds a great deal to it for me.

FÜ Thank you Martin. Conrad, do you want to introduce some other works in the show?

CS This one is called *Fracture* and you can see that it has the ghost of the form of *Paradigms* but actually it feels almost like it is shattered, like a light bulb that has exploded just before it falls to the ground. We have generated a helical stem inside the form which rises up that is quite easy to generate by just putting a perpendicular rod from the centre of every external face. It generates this stem through the middle and then from that we created the branching that holds all these leaves. There are about 350 unique leaves on each one of these pieces. This was an evolution of the works to make them more ephemeral, more ethereal. I wanted them to feel almost like a DNA model or like an empirical model, somehow it represents some strange amino acid, but there are very contradictory forces because within the model there is also this sense of entropy or change, there is expansion. So, it is a model of an amino acid that contains time. It contradicts itself and cannot be strictly defined. It can allude to these things but also other natural formations, trees and such like.

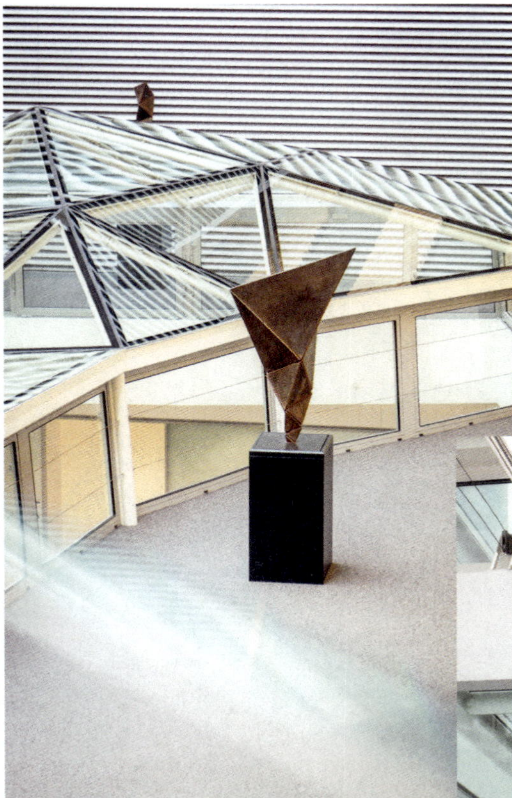

Paradigm Broad (Solid), 2016

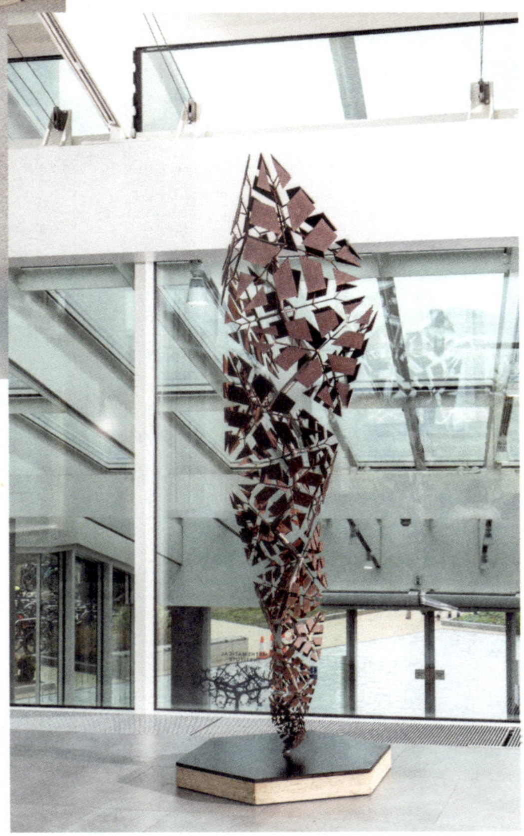

Fracture (R24W12), 2019

This is a memorial to the poet Yeats[9] in Bedford Park. We have used mirrored panels and it is all about the poem ['He Wishes for the] Cloths of Heaven'. I was trying to use materiality to connote the lines: 'The blue and the dim and the dark cloths/ Of night and light and the half light'. There were all these specific references but when you use mirrored panels it becomes very chameleonic, it disappears against the Autumn [sky], becomes brown in the sunset. It takes on the pinks of the sky in the summer, it is blue in the winter, it becomes grey, and it is always shifting and always changing and it camouflages itself against the weather. It is a dynamic piece and an experiment with materials.

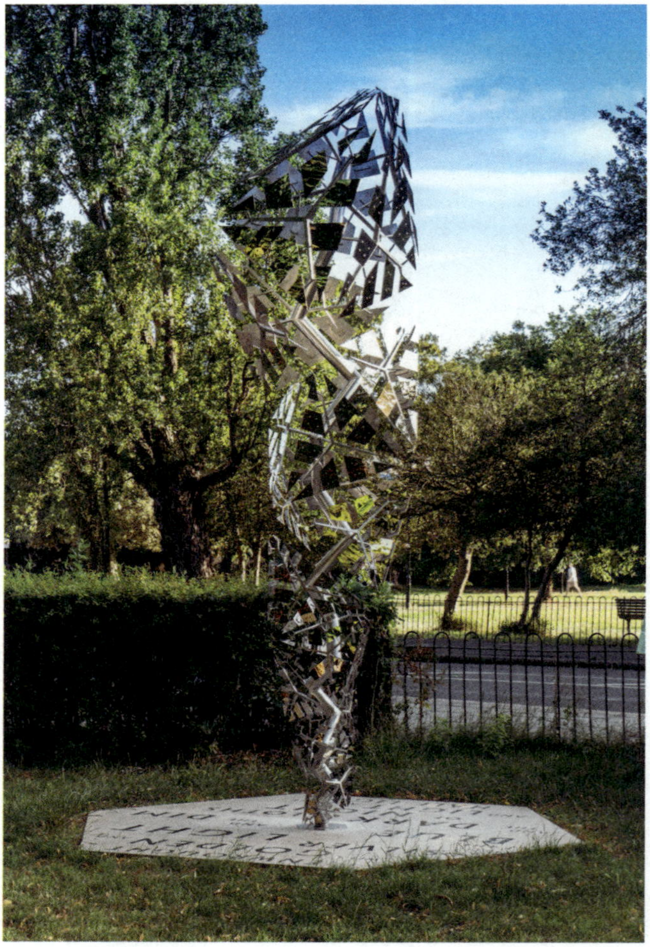

Fracture (R45S33) – Enwrought Light, 2021

These are the four *Beacons* which are studies for a piece I did in Ramsgate which children in Ramsgate chose. [Each *Beacon* consists of two optic, perforated discs which create interference that send a semaphoric message]. It says 'HOME' out to sea on the back and the front.

9. William Butler Yeats (1865–1939) was an Irish poet, writer, and dramatist. He was one of the foremost figures of 20th century literature.

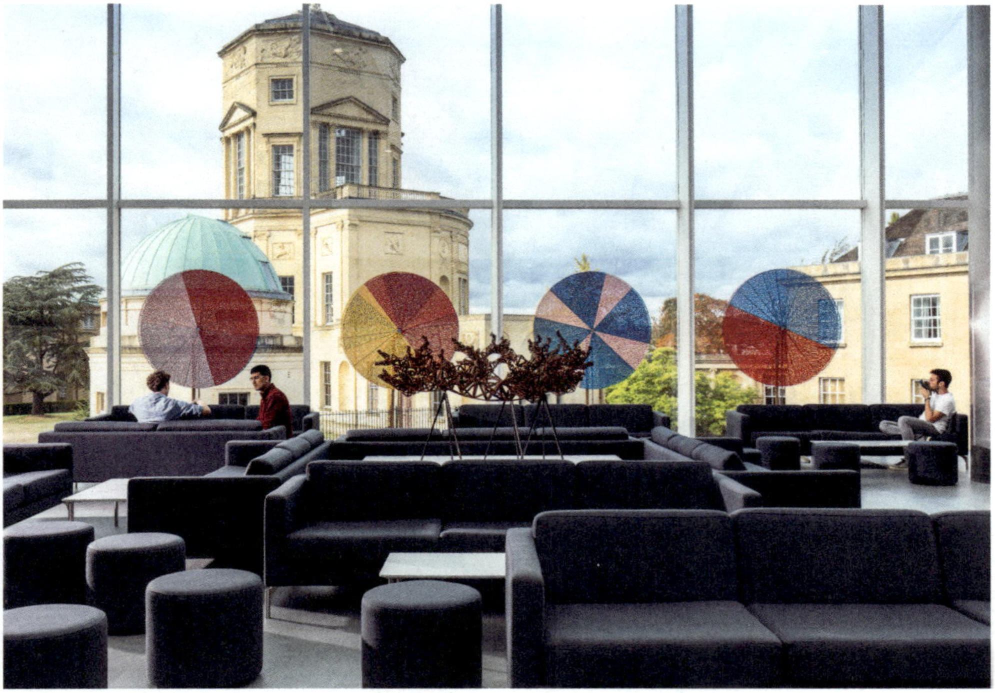

Beacons (he15D6), (om15D7), (mo15D8), (eh15D9), 2021

In front of that is the *Dappled Light of the Sun*, which is the evolution of the one you saw in that field at the beginning (my successful failure) where I have started to create generations of them, so that there is a cascading, generational [movement]. As the tetrahedrons emanate from the centre they cascade down in scale. This is the truncation sequence: you see the primary, the secondary, the tertiary, the quaternary tetrahedrons which seamlessly descend through these truncations. It creates this claw which [can be interpreted as] time moving inwards or moving outwards. Then again it has echoes of natural formations.

There were three of them outside Chatsworth House. This is the [central] work that you have in the common room. Again, it is an example of my engineer and I working very closely together. Each cloud sits on three tripods (so nine legs), but because of the proportions of the piece each of these clouds weighs five tonnes. But for me they feel almost like they are being tethered down because we have got the proportions right and they feel like they are ropes holding the work down despite their mass. If we had got those proportions wrong, they would have felt very heavy and clunky. We spent huge amounts of time developing the way the piece descends. It is all based on Fibonacci and the golden section,[10] the proportions of the edge size to the K rate, it is all scrutinised a 100 times to get to perfect proportions or attempt to get to a perfect proportion.

10. The golden section or golden ratio is a ratio between two numbers that equals approximately 1.618.

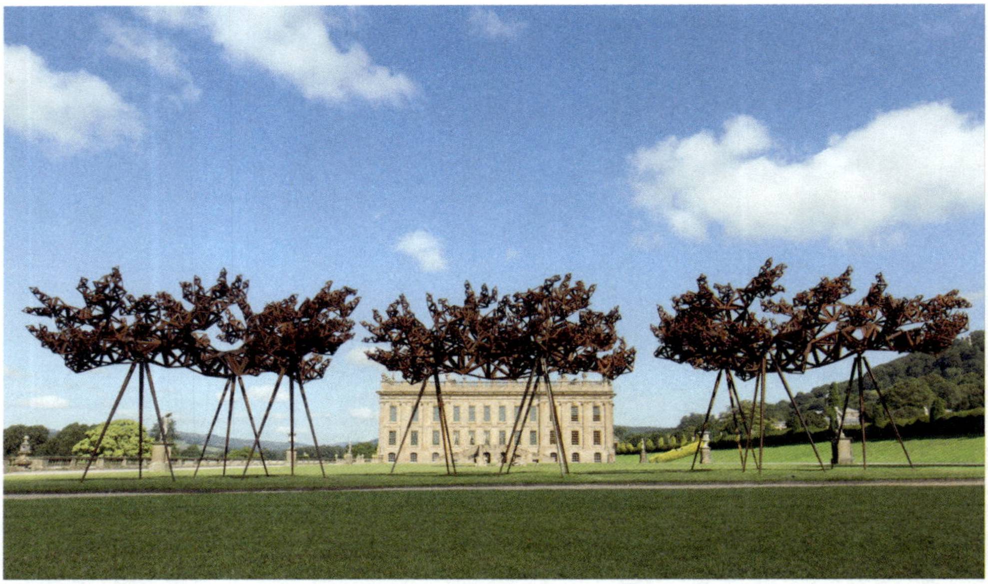

The Dappled Light of the Sun, 2015. Installation view at Chatsworth House.

Looking up, this was in the Royal Academy Courtyard. This term 'cascade' was something that I used with these pieces. It then informed this title; 'Cascading Principles' was a term that I liked because it implied this cascade of scale throughout a lot of the works but also this idea of something irrational. A cascading principle, maybe a collapsing principle.

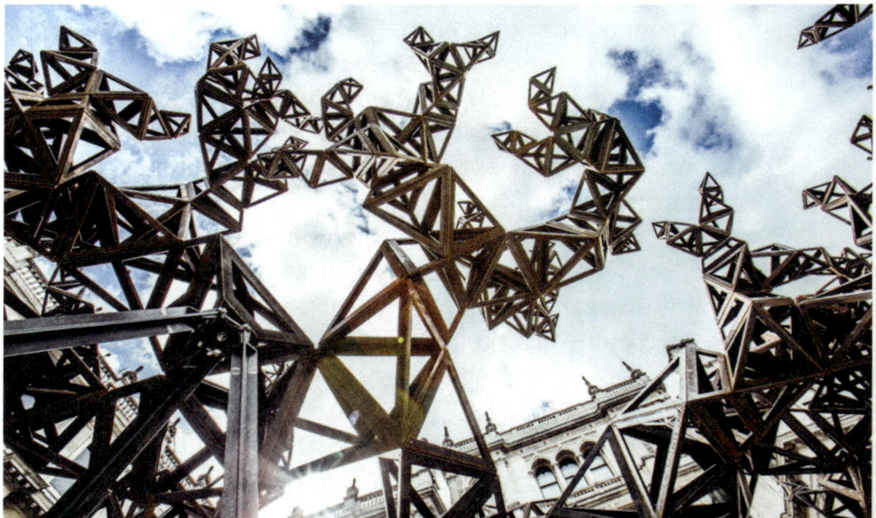

The Dappled Light of the Sun, 2015

FÜ Thank you Conrad. Martin, in a way you already revealed your favourite work in the exhibition. Besides *Schism* is there any work that you would like to live with, or something that you really enjoy seeing every day, or [that] makes you think of something different every day?

MB I find it [*Schism*] so provocative, it is a moment of break in the day. You do notice it, as I was saying to James yesterday, it takes my mind off the mundane administrative things. It does not necessarily give you a mathematical idea, but it makes you think it is a beautiful thing. It turns your mind back onto mathematics, I think. I get that trigger benefit from walking past any of the works, but it is *Schism* that, for the reasons I articulated earlier, resonates most deeply with me.

JS I agree with that. I also love the *Fracture*, it is just beautiful to me. Again, it is hard to say why you find something beautiful or elegant but I really do like it up in the common room.

I had visitors last week, collaborators who came over from London, and the first thing that they noticed of course were the sculptures and they immediately wanted to look around all of the sculptures. They were asking me about it and I said that we were doing this [lecture] this week. They were my collaborators so they were coming anyway, but I think the wonderful thing about the exhibition is [that] it is bringing people to the Maths Institute who would not ordinarily be here. That might be some of you here today or people watching online, and it is also an opportunity just to reflect, as Martin said. Even if a mathematician is just sitting and looking and thinking a little bit, reflecting on geometry and space. Certainly, as geometricians, that is what drives us: developing intuition in geometry. You are certainly a geometrician too, Conrad. You have amazing insights into the geometry of space in what you do with your work.

MB That I agree with a lot, but I also think we do try to bring a lot of young people into the building in particular and I think it is an invitation to geometry. Why are those forms chosen, why do they fit together like that, what is it about that growth that is special, that works, and so on? I think that aspect of having them as a natural unspoken invitation to mathematics in the building is very good, particularly for young people.

CS I think there is a parallel between art and mathematics (you can confirm or deny this); certainly the excitement for me about a good piece of art is that everyone sees something different in it. It is the breadth of interpretation of a piece of poetry that is exciting to me. If everyone in the room saw the piece on the wall and they saw the same thing, it is probably not a very good piece of art. It is the breadth of interpretation that is exciting.

FÜ Lets conclude our conversation with a work that is not included in the exhibition. A special feature for our attendees.

CS This piece is called *Slow Arc Inside a Cube*. I had been making a lot of works with light and movement and the shadows were always a by-product of that. I spent a lot of time in the maths department at the Science Museum when I was a student and when I was feeling low or uninspired. I spent an unusual amount of time in a room full of 1960s vitrines, with psychedelic wallpaper and incredible Victorian objects and contraptions. There was a quotation from Dorothy Hodgkin,[11] who pioneered this process called Crystal Radiography, and she described the process as like trying to work out the structure of a tree by only seeing its shadow.

I was really captivated by this idea. We have all sat underneath a tree with a bottle of wine and looked up at the dappled light of the sun. You can imagine the leaves and the structure to a point, but the bark and the ants and the ripples along the surface is impossible. With this piece it is a simple premise; I have made a number of versions of these over the years. It started off with a simple cube with one layer of mesh and a light traveling from one side to the other, but over the years I have made it more complex. This one has two lights and three layers of cubes but what is interesting is there is no smoke and mirrors. You see the cube, you see the layers, you see the light inside. In a way you have a God's eye view; you can see the real, which is the cube, and the shadows (maybe a Plato's Cave[12] analogy). We can only see the shadows, but as mathematician, if you look at these shadows, would you ever be able to work out that cube from the shadows?

MB In principle, yes.

CS As an artist and as a mathematician or as a scientist, it is what we are all trying to do. We are all trying to ingeniously see around corners, infer what is real through very limited information. We are trying to see beyond our perception envelopes [but] also the more we learn we realise how flat our perception envelope is. We can slide it under the door. It is that contradiction, as you talked about earlier. The more we know, the less we know, but this feels like a really good metaphor for the human condition and for the scientific situation. I think what we share in the science and the arts is trying to understand what is real through repetition and through experiment and through endeavour.

11. Dorothy Hodgkin (1910–1994) was a Nobel Prize-winning English chemist who advanced the technique of X-ray crystallography to determine the structure of biomolecules.

12. Plato's cave is an allegory presented by the Greek philosopher Plato (c. 428 – c. 348 BC) to compare the effect of education and the lack of it on our nature.

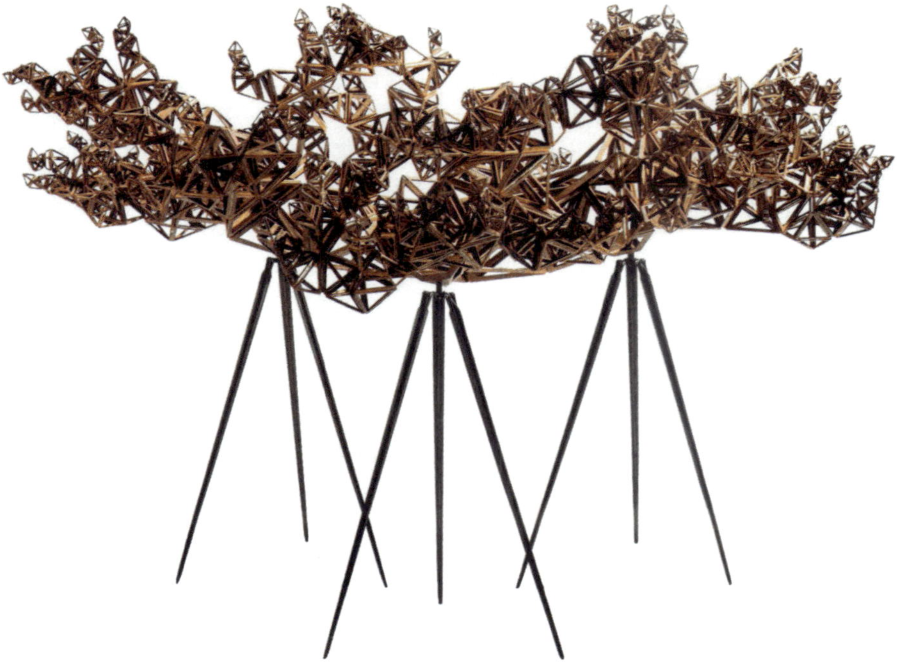

The Dappled Light of the Sun

The Dappled Light of the Sun is part of the *Formation* series, in which Shawcross took the tetrahedron as his brick, covering the space with steel 'tets', mated together in sprawling tendrils that cascade down in size as they dilate in density.

The artist's piece models the irregularities of this ostensibly basic form, the glitches and contradictions of its geometry. At first sight the piece can seem quite severe and brutal, but in its recollection of nature and the dappling effect of sun passing through its steel branches and leaves, it has a softness and a rich poetic and metaphysical potential.

Beauty in Mathematics and Music

Prof. James Sparks

As someone who was drawn to mathematics and music from an early age, the connections between the two have always fascinated me. At a fundamental level the elements of music are governed by mathematics. For example, certain combinations of notes sound 'harmonious' because of the mathematical relationship between the frequencies of the notes. Musical harmony, the subdivision of music into bars and beats, the different permutations and combinations of rhythms, and so on, all give music an inherent mathematical structure. In fact, just like mathematics, there is even a special notation used to describe that abstract structure. However, I think there are other, perhaps less obvious, connections. In a sense both mathematics and music are constrained, abstract, logical structures, but within these rigid constraints there is enormous freedom for creativity, with an important role played by both symmetry and beauty.

Mathematicians studying the foundations of mathematics are really studying structure, and the relationships between abstract structures. An equation A = B is of course a statement of a relationship, saying that A and B are equivalent, in whatever sense is intended. It is straightforward enough to start writing down true equations, but this is not what mathematicians do. Mathematicians seek interesting, elegant, or beautiful equations and structures. There is a strong aesthetic input. The way that mathematicians work, especially in the early stages of an idea, is often non-linear and intuitive, with more linear and methodical reasoning coming later. In music a composer often works in exactly the same way, but they do so for similar reasons: in both cases one is simultaneously trying to create and discover interesting and beautiful structures within a constrained system. Once you start to create, the constraints immediately lead to many consequences – sometimes wonderful consequences, but more often not what you are looking for – and one needs to use intuition to guide this simultaneous process of creation and exploration.

For some mathematicians, the connections between mathematical and musical creative processes extend further still. This was particularly true for Albert Einstein. Remarkably, he said the following about Relativity, his geometrical description of space, time and gravity:

> The theory of relativity occurred to me by intuition, and music is the driving force behind this intuition. My parents had me study the violin from the time I was six. My new discovery is the result of musical perception.

I would love to have been able to ask him more about what he meant by this! His wife Elsa once remarked:

> Music helps him when he is thinking about his theories. He goes to his study, comes back, strikes a few chords on the piano, jots something down, returns to his study.

I do the same when I am working at home and have always regarded it as mere procrastination, but perhaps there's something deeper going on. The aesthetics one is seeking in mathematics and theoretical physics are common also in music. I think Einstein was looking for simplicity, harmony, and beauty in his work, and music was for him an inspiration for this.

The notion of beauty in mathematics is hard to make precise, but for me one aspect of it has something to do with finding simplicity and complexity at the same time. By 'simple' here of course we do not mean trivial, but rather something natural and elegant; and the complexity is often initially hidden, to be uncovered by the mathematician. For example, take group theory, which is the study of symmetry in mathematics. The axioms of group theory are extremely simple, but it took hundreds of mathematicians more than a century to understand and classify the basic building blocks of these structures, which include extraordinarily complicated mathematical objects. To paraphrase the mathematician Richard Borcherds,[1] there is no obvious hint that anything like this level of complexity exists, hidden in the initial definition. This is the sort of thing that mathematicians find beautiful. Of course, symmetries and patterns play a central role in both mathematics and music, and this is perhaps another reason why so many people are attracted to both.

The combination of simplicity, complexity, symmetry, and beauty in music reaches a pinnacle in the compositions of Johann Sebastian Bach. Much of Bach's music makes use of counterpoint, where independent melodies are woven together. He often builds large, complex musical works, with many such simultaneous melodies, starting from only a small fragment of a theme. Bach then systematically works through different combinations and permutations, much like a mathematician might, making repeated use of symmetry and patterns. Writing music like this involves a great deal of analytical skill and is very similar to solving a mathematical problem. Starting with a small, simple idea, and creating/discovering a large structure from it is very appealing to mathematicians – it is elegant. It perhaps also inspired Einstein, who was a great admirer of Bach's music. Bach's genius meant that he was able to use this approach to create beautiful music that also has a more abstract mathematical beauty. For me, it is this combination that makes his music so special.

1. Richard Borcherds is a British mathematician currently working in quantum field theory. He is known for his work in lattices, group theory, and infinite-dimensional algebras, for which he was awarded the Fields Medal in 1998.

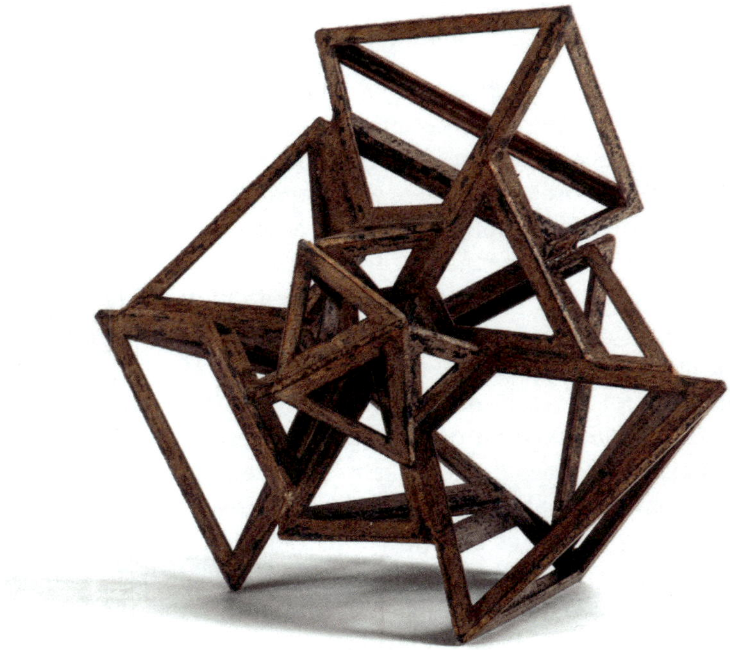

Perimeter Studies

This sequence of ten works, five structural and five solid, explores the geometry and spatial properties of the Platonic solids: the tetrahedron, cube, octahedron, dodecahedron and icosahedron. Shawcross explained that:

"I remember learning about the Platonic solids at primary school and was told there were only five and they each had a certain number of sides. It was not until much later, when I read a different definition, that they became a revelation to me. These solids are so important because they are the only way to divide the surface of a sphere equally. Once in possession of this knowledge, I was then able to exploit the radiant properties of the forms, pushing, pulling, and extruding the faces out from the centre."

The works explore notions of expansion, entropy, growth, and retraction. Each is a universe in itself, seen from beyond its horizon. These structural and solid arrays allow their properties and geometries to be compared, contrasted and defined against each other, like musical notes within a score.

Lecture Two, 27th July 2023: Envisioning Imagination

Speakers: Sir Roger Penrose, Carlo Rovelli, Conrad Shawcross
Moderator: Fatoş Üstek

Fatoş Üstek
Roger, Carlo, and Conrad, I am truly delighted to be exploring the depths of imagination with you and would like to invite you to draw out as many personal examples as you feel comfortable sharing with us. My first question is about the fact that you are all invested in other disciplines. For instance, Conrad, you derive inspiration from geometry and physics and your sculptures also necessitate a foundational knowledge of how matter responds (such as steel or wood) and how gravity works. How does being informed by other forms of knowledge production inform your practice?

Conrad Shawcross
Thank you, Fatoş. So, I think it all started actually at Oxford. I was at the Ruskin and it was an autonomous art school, and still is, within the university. I was suddenly plunged into the fabric of the university and all my friends were scientists and English students and poets and historians, and the course was also very much outward looking. It was not an internal course just studying the history of art and making in-jokes about [it], it was very much about exploring the grand history of ideas. I owed a lot to my tutor, Brian Catling,[1] who was a real inspiration at the time. I spent a very unusual amount of time in the Science Museum's maths gallery whenever I needed inspiration. I would probably go there on a weekly basis. It was this very 1960s view of a lot of Victorian mathematics, but I would pore over the artefacts and the objects, [their geometries were] a huge inspiration. I have just been doing this wonderful residency in Nottingham for ten days, studying primordial black holes, which is very good timing [for this talk]. I have been up there studying black hole ring down and gravitational waves and it has been hugely inspiring. I have always found inspiration in things beyond art and then try [to] fuse that into my work.

FÜ Thank you. Carlo, I want to ask you next actually; you have a gorgeous choreography of interest in literature, philosophy, and scientific thought. Can you please expand on why you think these disciplines are such good companions?

Carlo Rovelli
Let me put it this way. I think there are two reasons: one is that the way we think about the world, about reality, about ourselves, comes from this culture which is made by [different] components:

1. Brian Catling (1948–2022) was a British sculptor, poet, novelist, film maker, and performance artist.

history, science, geography, politics, morality, literature, art. They make sense if they are integrated (which does not mean reduced); they are different perspectives on the same thing, which is around us and [in] ourselves.

I have always been curious in my life, I guess in the same way as Conrad. In my books, I make an effort to talk about this integration. I am fascinated by science, by physics (by black holes specifically), but I think that to make sense, both for me and for all the other people, this [needs to] come into our worldview and inform our old view. Because this worldview keeps changing historically, not only [over] centuries, but I would say decades after decades, sometimes years after years, so it is a continuous conversation between the various species and the various parts. I think this is the best way of seeing what is going on in science, in art.

The second reason I think is that there is not only a different perspective on reality for all of us, but they very much influence one another continuously. I think that you know Roger was in contact with Escher in [the past]; I cannot believe this has not had an impact on his way of thinking about spaces, spacetime, and geometry. It happens all the time. If you read Milton there is Newton, then Copernicus, cosmologists coming in. If you look at what happened in quantum mechanics which was happening in the same years [that] Braque and Picasso [were] breaking up reality, perspectives. I grew up in quantum mechanics. Reality has different perspectives that come exactly from this combination of things. I see culture as this continuous exchange: different aspects of it nourish others, of course each one [has] its own tools, aims, language, and so on.

FÜ Thank you. It is almost like there is a lot of correlation between disciplines, but the manifestation might be different because of the tools that they use or the areas that they look at. Roger, you have worked on two different and very closely related disciplines: mathematics and physics. So, you can operate between two knowledge fields [and] you are also interested in the arts. We will talk about the relationship that Carlo mentioned, with Maurits Escher,[2] later on. I want to refer to your relationship with literature. You bookended your book *Cycles of Time* with a short story, and collaborated with author Brian Aldiss[3] on a science fiction novel titled *White Mars*. What does working within and outside the frameworks of your field bring to your practice as a mathematician physicist, physicist mathematician?

2. Mauritis Escher (1898–1972) was a Dutch graphic artist whose work was often inspired by mathematics.

3. Brian Aldiss (1925–2017) was an English writer and artist known for his science fiction novels and short stories.

Roger Penrose

Well, it is difficult to say. I should explain first that I have a strong inheritance in the arts in the sense that my [paternal] grandfather was a professional artist. [His name was] James Doyle Penrose; he was Irish and he was [from a] very strong Quaker family on that side. But you could see he was very pictorial; he only had wild things but they were all very religious [in their] topics. My father very much inherited the artistic side, as did my uncle Roland. He became one of the surrealists; he knew Picasso and many of the others. He was in with the group there and became a leading surrealist in this country. I think his father would not have approved of his artistic activity. My father was a good artist but he was not in the professional artistic side. He was very good at drawing and painting and he used to sketch things. He would go out sometimes and I would join him and do the same but it was [a] very direct pictorial representation of what he saw. I used to draw all sorts of wild things in my notebooks, completely different, that he would not have approved of, I am sure. Absolutely wild distorted human features and so on all over the place. I think he would not have liked it at all but he did not see them.

I think I am certainly a very visual person. When I was an undergraduate I was supposed to be doing medicine (both my parents are medical) but I disappointed them by going [for] mathematics. This was at University College London and they had a rather unusual system. It was a three-year course, but at the end of two years you did most of the papers and in the third year you worked on your specialist topic. Being somebody who was very visual thinking, I noticed that visual thinking was rather a minority thing in mathematicians. Most people did not think that way. In fact, it was probably only about two or three people in the club apart from myself who did tend to prefer thinking visually. I decided to do two geometrical topics for my specialist paper and I ended up getting a first, but a lot later. My best paper was not in geometry, it was on algebra, and this was because I never really finished the geometrical papers. I would see how to do the problem, then I would have to translate that into words, write the words down, [with] the translation process going on. I was a slow writer as well; I just simply did not finish the papers. I [started] to think that there was a selective effect against being too visual in mathematics and certainly when I taught mathematics, I found that the people who think visually are rather in a minority. I suspect the reason is more or less what happened to me in a way, is that it is harder to test. The examinations are much more geared to being easy to do if you are not such a visual thinker, so that was a little disappointing to me. The really good ones are good at both, certainly that was true, but in my case, it was very much visual thinking which was the main thing, and that related to books that I wrote and so on. I like to do all the pictures myself.

CR Roger probably remembers this: I had the great privilege of being in Pittsburgh, sitting in a corner with Roger and Ted Newman, a great mathematical physicist who wrote the general solution of black holes. We were working together, so when Roger was coming, I just came in and sat in a corner. It was absolutely fantastic, because the two seemed not to understand each other at all at the beginning. They would go for a while and then [at] some points [they would say]: 'Ah, yeah maybe' and go to the blackboard and say 'totally' and then they were understanding one another; I think science is also this.

RP It was a very good collaboration because we thought so differently in a way. He had a wonderful skill with having a feeling for equations which I never really had at all.

CR Manipulating it, turning them around; incredible.

FÜ I want to bring in [the] artist Cecilia Vicuña. She has a very strong visceral relationship with words as well. She was commissioned for the Tate Turbine Hall in 2022. She breaks imagination into two words, for Spanish, then in English, and suggests that imagination is 'image-in-action'.[4] Maybe, Roger, we could come to your beautiful drawing of a mermaid that you drew on the occasion of your lecture at the Hans Christian Anderson Academy? The drawing illustrates the quantum and the classical levels of reality, so can you perhaps talk us through [this]?

RP This is the Little Mermaid, based on the Hans Christian Andersen story. The bottom part of this picture represents what I call the quantum world, where you have all these strange things which are not familiar to people (it has got these strange fish and hammerhead sharks and things like that swimming around, various unfamiliar things) and the top of the picture is more the world we are familiar with. That is the classical world. And the line across the middle integrates the difference between the two. What is the mermaid doing? Well, you see, she inhabits both worlds, in a sense. She is able somehow to bring the quantum world up [to] the classical world and she represents in a certain sense the big paradox in quantum mechanics, which is the thing called the 'collapse of the wave function'.

You have a formalism in quantum mechanics; there is a thing called the Schrödinger equation and that tells you how the quantum state evolves in time. It just chugs away and follows this rolling equation except that it does not; [at] a certain point you have to wheel out of the cupboard a measuring device and then this performs an experiment. You get certain probabilities out and you wheel it back into the cupboard again, but that is not following the

4. Cecilia Vicuña (b. 1978) is a Chilean poet and artist. Her work is noted for themes of language, memory, dissolution, extinction, and exile.

Schrödinger equation. It is a big mystery how is it that every now and again you have to cheat and change the rules and then change back again to where you were before. It always struck me that quantum mechanics does not really make sense and you have to accept that it does not make sense. I think the trouble is that so many people in the history of the subject tried to make it make sense as it was, whereas it seems to be clear that without a really serious change at some point in the subject you are not going to make it make sense. So, this is really illustrating [that] you need something else which is what the mermaid is doing in this picture, she is something that happens in both worlds. She brings one world up into the other one in a way which is mysterious. As far as we know, when we understand the physics properly it will cease to be mysterious, but at the moment it is not explained by any theory that we have.

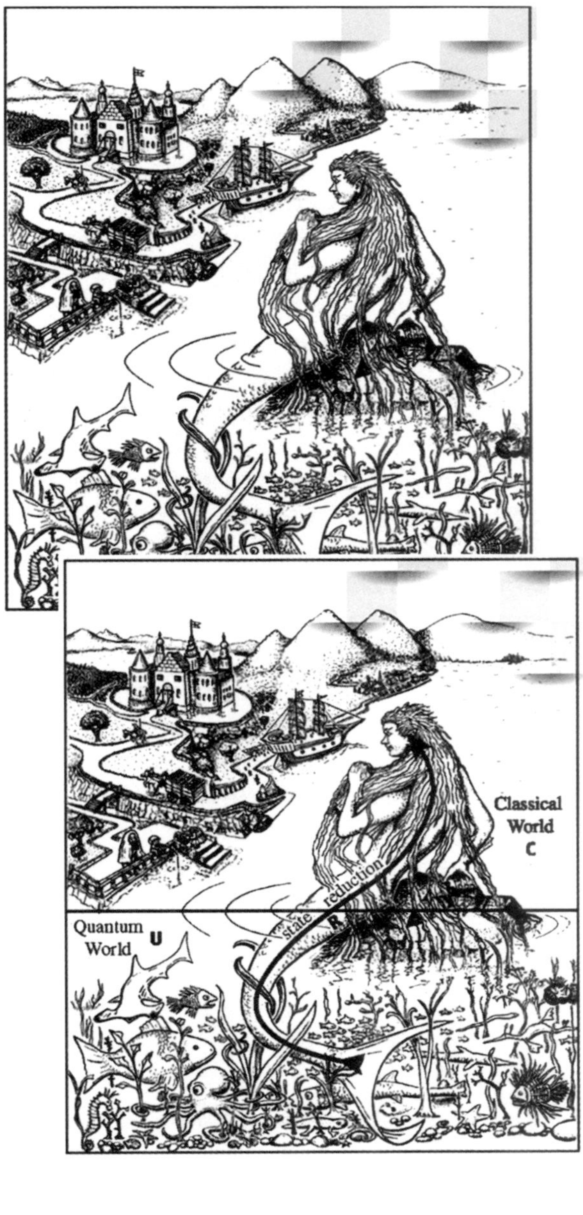

FÜ Thank you. I will come to you Carlo on this one, because this paradox is quite interesting if you think about quantum mechanics. For instance, this morning we were talking about particles that act like waves and sometimes act like particles (let us say light). Then how can we imagine something that is so paradoxical as a particle and a wave [simultaneously]? I could not ask the question properly...

CR You did.

FÜ Perfect, I did great.

CR We are all still fighting about that; we have not got any consensus. Everybody has their own ways about it. I have my own which of course is [the] right one and everybody else [has] the wrong one...

FÜ What is yours?

CR Others have also the same idea that they have the right way, but that is the beauty of quantum mechanics: its incredible effectiveness in describing the world and giving us tools for all this technology, allowing us to understand how the sun works, how galaxies were formed, why matter is stable. But then, when we try to make sense of it, we clash against this difficulty that Roger was describing, [to] come up with an actual understanding of what is going on. We have not yet [understood], or perhaps we have various ideas that something else different can come in, that we have to look at the world in completely different ways, but so far nobody has been able to convince everybody else and that is the beauty.

We do not have to live [in a] finished world. The world is this incredibly kaleidoscopic jumping around of events around us, which we [think] we understand because we get used to it, so we sleep walk through this immensity, this beauty, this strangeness, without really realising how much we do not understand.

To me it seems that the best science (and the best art) is when it shows the breaks in our sleep-walking, close-minded belief that we understand reality and there is nothing strange. There are these moments in which you see things differently because the progress of science leads to Einstein, or because we understand that the Earth is rotating so we are not moving, we are just turning around, or because you look at Vermeer[5] and say oh the light is different, you read Shakespeare and you rethink humans differently.

FÜ Fabulous.

5. Johannes Vermeer (1632–1675) was a Dutch Baroque period painter who is perhaps most famous for his *Girl with a Pearl Earring*, painted in 1665.

CR This is a moment in which art and science [work] in totally different ways but give us new eyes for looking at the world.

CS When I was a child, my first inkling was that reality was not as it seems. It was provoked and jolted through science and mathematics and that set me on that pathway. I have always been excited by the rug being pulled out from underneath me and with my work that is always about the challenge of reality and trying to erode this sense of certainty. Science does that all the time and maths does all the time, it is why it is exciting.

FÜ Perhaps it is a perfect time to introduce your *Slow Arc*, Conrad?

CS Okay, sure.

FÜ Buckle up everyone. We will now turn off the lights and take you for an experience. Please do not move from your seat.

CS This piece was inspired [by] the Science Museum in London. I spent a lot of time there and I [did] a residency there and I have made a lot of kinetic works which are a by-product of those times. They were moving lightworks but I would always ignore the shadows, this is a by-product, and then I came across this beautiful Dorothy Hodgkin quote where she described the process of radiography as like trying to work out the structure of a tree by only seeing its shadow. It was a beautiful, humble quotation.

This is called *Slow Arc Inside a Cube* and it is probably the 11th one I have made. This one has three cages and two lights which move in the sinusoidal way, so the lights accelerate and decelerate. The first one had one cage and one light and it was much simpler, but still had complexity. What is exciting for these pieces for me is that, quite quickly by the layering, you realise how close complexity or chaos is. You can look at the ceiling and the floors and the wall and you guys are all illuminated with the rhombic tiling but you will get almost the liquidity of water on one side and then you get this spatial dimension. I like it also in the context of physics in that it almost obliterates the present, everything is being pulled and twisted and stretched in every possible direction and it will never stop.

But I also love this entity in itself, this sort of Plato's Cave analogy where you have got the shadows of the real. The metaphor is that we are the scientists, the artists, we can only see the shadows, but as the viewer you can see the cage, the platonic, and the shadows, so you are able to see both. You are put in this sort of God's eye view of being able to see both platonic and the shadows of the real so it is quite a good, useful mechanism for showing the condition of the artist or the human condition where we are ingeniously trying to see beyond our perception envelope. I think I will leave it there – we can turn it on later.

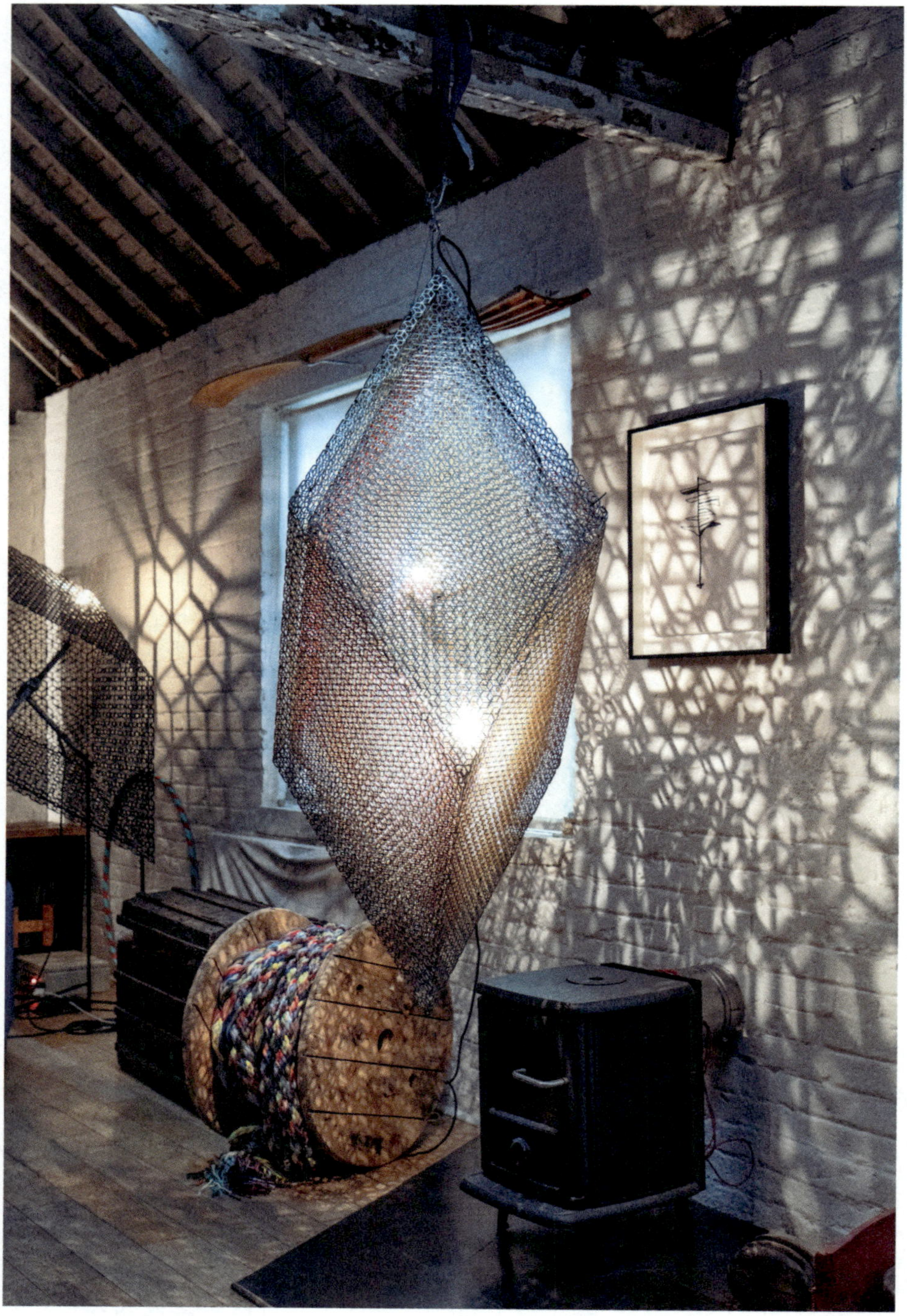

Slow Arc Inside a Cube XI, 2020

FÜ That was very beautiful, thank you. Do you have any immediate impressions, Roger and Carlo?

CR Models are not any more real than the thing [itself].

CS Well, that is a debate, yeah. I mean the shadows of the shadow: you can have this kind of infinite recession within it.

FÜ Perhaps we can continue with the platonic transmigration, the graph that Roger has beautifully drawn. I want to ask all of you if you have a methodology of imagination when you think about complex ideas and bring them to some form of clarity. I personally find the illustration of the platonic transmigration utterly inspiring because the drawing so immediately conveys that smaller part of the mathematical world and compasses the majority of the physical world. I think it really coincides with the [experience of the] *Slow Arc* as well. Roger, would you like to introduce us to what your thinking was behind this quite intricate diagram?

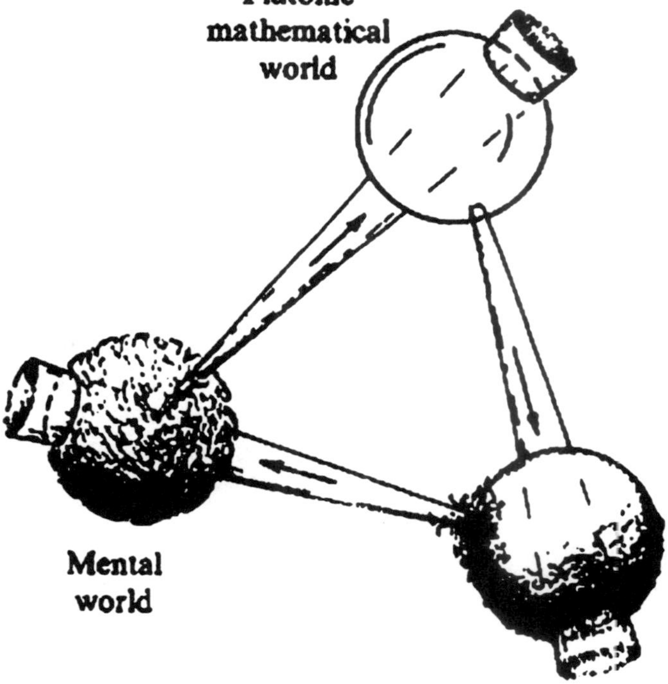

RP I refer to this as three worlds and three mysteries, and the worlds are the big sphere things and the mysteries are the connection between the two. So, the top thing represents mathematics and I regard this as having a reality independent of ourselves. I think people [who] are not mathematicians do not quite have the same feeling. There is a lot of feeling a little bit like archaeolog[ists] in mathematics, you are digging away in this world and you are finding something – a little wonderful crystal or something like that – and it is all there already. It is not as though we are creating it, this definite feeling is that it is something out there. So that is the top world, this mathematical world. It has an existence independent of ourselves in a sense, but then it is only a very small part of that world [that] really directly applies to physics, if you know about mathematics. We have some wonderful equations which do apply to the physical [world] (they apply extraordinarily precisely). It is remarkable, when you get them right, how well these laws describe the physical world, but it is a very small part of that. Then again you see there is the mystery, why is it that somehow the physical world depends so beautifully on these little small parts of mathematics. Then you have another world, which is the physical world down there (actual things like this chair and all that stuff), but only a small part of this physical world actually has conscious experience. That is the world down on the bottom left.

I have got one that is left, that is the mental world, which is conscious experience and understanding, qualities that we use our mentality for. Nevertheless, we are built out of stuff which is the same kind of stuff (as far as we know) as everything else in the physical world. It is just organized in some clever way which brings [an]other thing out of it again. It is only [a] very small part of the physical world which has mentality. Most mental frameworks have got nothing to do with mathematics; you can walk down the streets and look at the cars go. Of course, if you ask about the dynamics of how cars work, that is going back a little bit but you can appreciate things without that. You can look at the beautiful scenery, you can appreciate the sky and the stars in the sky, or beauty in the flower or in the bees or animals and things like that, and you are appreciating them at a quite different level. You do not need to know what laws govern them and so it is a very, very small part of consciousness which is actually concerned with the mathematics. Even mathematicians do not spend the whole time thinking about mathematics; some of them probably spend most of their time, [but] not the whole time. So, it is a very small part of mentality which is concerned with this platonic mathematical world. They explore all parts of that world, it is just nothing to do with a little tiny bit which is concerned with the physical world. So, it is described as a kind of paradox. It is drawn as a paradox also because it is an impossible triangle; each bit of it is coming out of the element and coming towards us, you could not construct that out of out of rods and fear of things so it is meant to be a deep puzzle beyond all the little puzzles which are involved in the whole picture.

FÜ But I think what is striking is that it is a very simple diagram, but at the same time you can write volumes about what this diagram would mean and what it would correspond to in all worlds.

CS Its implications are massive.

CR Roger, do you think that in the platonic world there is more than mathematics?

RP No, I mean it depends what you mean by mathematics. It is mentally mathematics, and it is only a small part of it which has anything [to do with it]; they would have a certain equation (Einstein's equations or Newton's) nowadays, and a few things which have to do [with it], things that are neither of these three worlds.

CR So do you think that there are [things] outside these three worlds?

RP I mean that is the mathematical, that is what that sphere represents. I am not saying there is mathematics outside mathematics because it is supposed to be mathematics.

CR No, but other platonic things which are not mathematics?

RP Oh, I see, well that is a different story. No, I am talking about the platonic mathematics there. You can consider [that] other things have platonic existence but then maybe you follow them down into one of those other worlds. In a sense the whole picture was, you might say, [that] the mental world has this platonic aspect.

CR Morality has its own platonic world or...

RP No, I agree with that. Yes, morality is not up there.

FÜ Carlo, let us explore a drawing by you this time. It is more a graph than a drawing that you have made. It is from the lecture that you gave in India. I believe it is where you met Roger, in the beginning of the 2000s. This graph is an outcome of your research about scientific papers and the topics that the scientific papers were mostly addressing in a specific period. My question is not necessarily the explanation of this graph in particular, but rather a general question about data. Does data present you an interesting creative platform to build your ideas?

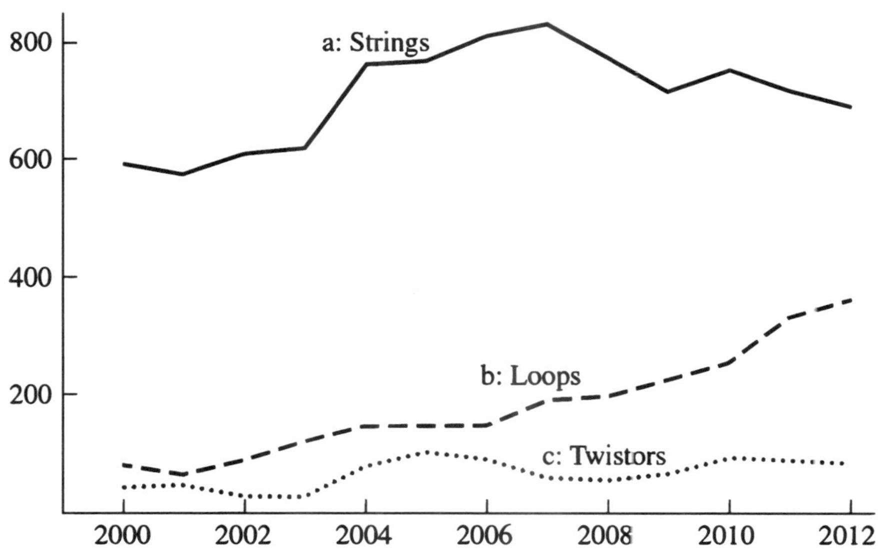

CR Well, this is a proof that fashion has nothing to do with good science.

RP That is the number of people who think about these different topics.

CR This is the number of papers published in Europe [on] these three topics [String Theory, Loop Quantum Gravity, Twister Theory].

RP So, you see string theory was way up there.

CR At the time, now it is much lower.

RP At that time (I hope there are not so many [people working on it] now) the one at the top was the majority point of view and then Carlo's one was the middle line. That is loop quantum gravity as far as I can remember and then I am sort of creeping around at the bottom – that is twister theory at the bottom.

CR Data is also the thing [to] nourish science, but science is not just looking at data and taking things out of it, it is [a] much more complicated operation, not only because of that.

RP But also, because the fashions in the world [are] not so much about truth.

CR Yes, fashion is not about truth. Roger has a book in which [he] discusses this.

CS It is also about the imagination of things. There are so many ways you can interpret and represent data and information but [the] key [is] in terms of where the human element comes in.

CR Science builds on data, even a super abstract thinker like Roger thinks about aeons of universes repeating themselves and then looks for data in the distribution of things that come from the sky that supports this and tries to interpret it. Data needs to be interpreted and the way we interpret it [is] not clear. Let me give an example which I think is a major example. Mathematical science was defined for several centuries as the following: we see things moving in the sky (Mars, Venus, planets, the Sun, the Moon) [and] find simple movements that make sense of this strange movement. Mass goes back and forth between the stars so [there was a need to] find circular movements that reproduce this movement. Astronomers played this game for centuries, including Copernicus.[6] Kepler[7] changed the game and said instead: '[let us] find a combination of circles that make sense of what we see'. For Copernicus what we see is because we

6. Nicolaus Copernicus (1473–1543) was a Renaissance polymath who formulated a model of the universe that placed the sun at its centre.

7. Johannes Kepler (1671–1630) was a German astronomer, mathematician, astrologer, natural philosopher, and writer on music.

are ourselves, are emotional, and for Kepler we forget the circles and find an arbitrary mathematical curve, but it is not circ[ular]. This is an example in which the problem was completely redefined; at different levels it happens all the time. I know very little [but] art has been redefined [in] every generation. What are we trying to do when we do art?

CS Do you think that Copernicus, when he had this radical paradigm shift where things swap around, do you think it was because he saw in the numbers that we are in fact revolving around the sun, which was an unimaginable idea at the time? Or do you think he just had this radical image?

CR He was Polish, he came to study in Italy during the Renaissance. His friends were some of the artists and the thinkers of the early Renaissance. It was this strongly cultural movement of 'let us do everything' and he was deeply influenced by the culture around him to do everything. It was an idea that he found in ancient literature. It is a combination of the courage of thinking completely differently and a deep knowledge – finding ideas and recognising a good one.

Kepler is the opposite and again there is no methodology. It is one of the most incredible stories, because he was using the Copernican system with circles. He was working on Mars and Mars moves according to the Copernican circle theory as a circle of a circle (an epicycle). Except that he did not have computers so he had to do everything by hand, computing the movement and checking with the data. He decided that the circle of a circle could be approximated by ellipses. Ellipsis is easy mathematics to work with, easier than two circles one on top of the other. He was using the ellipses as an approximation of what [he] thought was a real thing, which was [a] circle of a circle, and then at some point in checking the numbers he realised that when he was doing the correction from the ellipses to the circle of a circle, he was going away from the data instead of toward the data. What I consider an approximation works better than what I think is the actual thing. Maybe I should just keep the ellipses. It is an example of the infinite amount of human stupidity. Nature had to put the ellipses in front of his nose, he is already using it before he could actually see it in the data; he [did] not see the ellipses in the data until the correction away from the ellipses.

FÜ This is a perfect segue to proceed to our next slide, which connects circles with the circle in a circle; this is an illustration of loop quantum gravity. It is not Carlo's drawing but perhaps, following on from Kepler, could you perhaps talk us through what loop quantum gravity is?

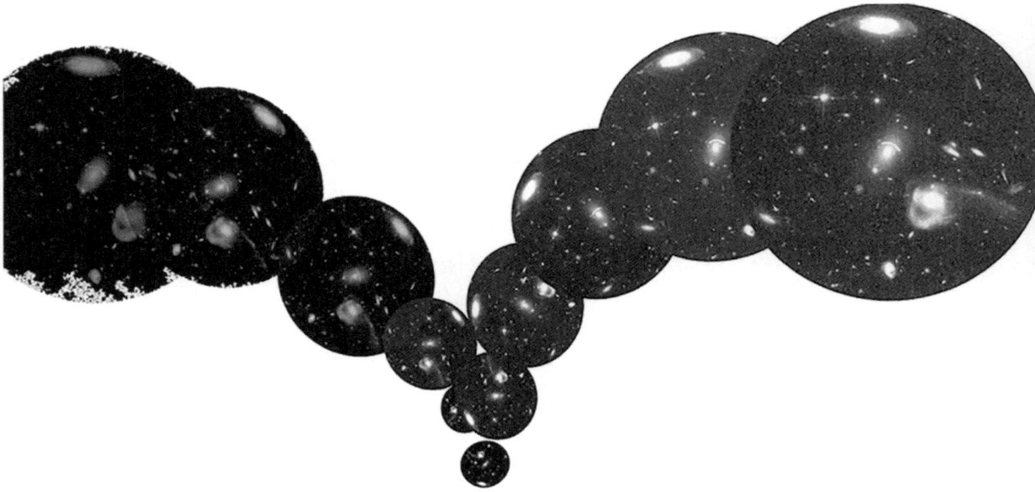

The big bounce of the universe by Francesca Vidotto

CR In my answer I have got to connect to Roger's idea twice. The first is that this is a picture of one of the predictions or consequences that seems to be coming from blueprint to gravity [in] the equations. When you apply the theory to cosmology and if you go back in time you start from the bottom here (this is a big universe), you go back in time and we all know that [the] universe in the past was smaller. What the theory seems to predict [is] that you keep going back in time; you do not just disappear in a point but you have a previous phase where there is a contracting universe. What happened at Big Bang is all bounce. This is a possibility that seemed to be indicated by loop quantum gravity. There are other possibilities. One is that the Universe emerged from nothing and one is a possibility that Roger has been studying a lot recently, that there is a previous universe. But it does not need to contract to rebound, there is a different kind of mechanism from which it expands a lot and then in some sense we start again. This picture [is] about trying to use what we know about nature today to understand what happened at [the] Big Bang and before and maybe for the future. The theory we used for this is loop quantum gravity.

That is my second connection to Roger. Roger, a zillion years ago, had pulled out of the sky a complete intuition about possible mathematics for describing quantum spacetime and he called it 'spin networks'. There was a mathematical structure called a 'spin network'. This is formed by a network (little links connected to one another). The links carry labels called 'spins' because the mathematics that describes them has to do with rotations, or 'spinning'. You attach these 'spin' labels to the links. This structure turned out to be precisely the mathematical tool we needed, years later, Lee Smolin and I, while working on making sense of quantum gravity mathematically. We found Roger's spin networks were the right mathematics to apply for the problem we were working on. Technically these spin networks define a basis in the space of the states of the theory. This gives the intuition of the description of spacetime that comes from loop quantum gravity. If you look at [a] very small space, you have got [the] smallest one, smaller smallest one, smaller, much smaller than atoms, much smaller [than] nuclei. If the theory is right, what we expect is that we see something similar to the mathematics of Roger, namely a discrete structure, little nodes, quantile space, connected by links. This discreteness of space is the core of the visual intuition of loop quantum gravity and the subtle point is that this network, you should not imagine it in a space with a network inside; forget the space, there is only the network, and what we call space is sort of the neck (seen from a large distance) of a t-shirt. It is continuous only if you do not look at the details. This granular structure of space is predicted by loop quantum gravity and is the main intuition about reality that comes from quantum gravity and the math comes from that guy there [Roger].

FÜ That is amazing. Perhaps, Conrad, it is a perfect time for your question for Roger.

CS I have been thinking very hard about the one question you would ask Roger Penrose. It is sort of a fairy tale moment. I imagine that you came across black holes eighty years ago as a young boy. You probably for the first time encountered the idea of them as a notion, as this vague theoretical idea and it is probably safe to say no living person understands black holes better than you do. The one question I wanted to ask you is about the future. It is about imagination, about what you think will be revealed. There is so much in our ability (through interferometry and all these amazing techniques) to observe collisions and a ring down. What do you think the next fifty years will reveal about black holes and how should we as laymen respond to these entities? These extraordinary, powerful, destructive [entities] for me they fill me with hope, it is like discovering a new world. But what does the future hold, what do you predict will be discovered?

RP May I first slightly correct what you are saying about these things? It was when I was in my second year as a research fellow at Cambridge, St John's College. I had a colleague who taught me an awful lot of physics. I learned mathematics, and Dennis Sciama[8] was a very good friend of mine and I learned an enormous amount of physics from him. He heard about a lecture that was being given at King's College in London and he drove me to this lecture which was being given by a person called David Finkelstein. He described basically what we understand by a black hole, in the sense that people previously had been very confused about this thing that is referred [to] now as a horizon. You have this region which you could fall into but you cannot get out again so it is a one-way barrier if you like.

8. Dennis Sciama (1926–1999) was an English physicist who played a major role in developing British physics after the Second World War.

You can fall through in one direction but not the other and it really had not been understood properly by people, I gather, until David Finkelstein. I thought other people like Lemaître[9] and so on had but apparently not really according to Dennis Lemkuhl who is a historian on these things; he tells me really it was Finkelstein who first understood it. It was a real revelation to me and I began thinking gosh I wonder if there is a theorem which shows that you cannot avoid these things. This was a long time before I found a theorem of that sort. That would have been...

CR '58

FÜ You mean 1958?

RP The Finkelstein Lecture? Yes, that would have been '58 I think. I began thinking, I wonder if there is a theorem that shows you cannot get rid of this horizon, you cannot get rid of the singularity in the middle, you have still got this place where things go wild right in the middle. It was not at what people thought was the singularity originally, that has gone into the middle, so I began to think is that one in the middle something you cannot avoid? Then I started thinking well I wonder if there is a theorem I have never heard of, what do I know that other people do not know in this area? Ah! Two component spinners. This was something I heard about from Dirac,[10] the great physicist. I went to his lectures with nothing to do with what I was supposed to be doing. I think there is a moral there. As a graduate student I went to three lecture courses which [were] absolutely nothing to do with what I was supposed to [be doing]. One was by Hermann Bondi[11] on general relativity which had a big influence on me. One was Paul Dirac on quantum mechanics. I learned about quantum superpositions and that sort of thing and I was very puzzled about it and then the third one was [a] lecture course by a man called Steen on mathematical logic and I learned about Gödel's theorem and how you could see the truth of certain things that your methods of proof would not get you to. I found that absolutely stunning. I do not know what the moral is to all this is but there certainly were worse things that were nothing to do with what I was supposed to be doing. I did write a thesis which was more or less what I was supposed to be doing but these other lectures really had a bigger influence on me. Certainly, the Finkelstein lecture which was just a seminar.

9. Georges Lemaître (1894–1966) was a Belgian Catholic priest, theoretical physicist, mathematician, and astronomer.

10. Paul Dirac (1902–1984) was an English theoretical physicist who is considered one of the founders of quantum mechanics and quantum electrodynamics.

11. Hermann Bondi (1919–2005) was an Austrian-British mathematician and cosmologist. He was best known for developing the steady state model of the universe as an alternative to the Big Bang Theory.

CS And the future? If you dare to hope for what we will find.

RP Well, it is not the future, you see, because it is already written up in papers. I have had this theorem about singularities in black holes (which was in 1964); I gave a lecture in King's College London, same college as the Finkelstein one, much later. The paper came out in 1965, that was the paper which I suppose eventually got a Nobel Prize. It took a little while but still, I have heard of longer. This tells you that, in gravitational collapse, which is what happens when you make black holes, you have these singular places where the theory goes wild, but fortunately it goes crazy behind these horizons so you do not worry too much about them. There is also a place where the universe apparently went crazy by itself, which was the Big Bang. Stephen Hawking picked up on these techniques I was using and he started to prove theorems which had to do with the Big Bang. Are they generic, can you get away with it by making them more general? Maybe there was a collapse which swished around and came out again and those results showed no, that does not work either. So, you have singularities, which you cannot get rid of. They seem to be the same thing but they are not, they are utterly different. The ones in black holes have an utterly different structure from the Big Bang and it took a long time for me to appreciate this.

I remember not realising it for a long time and I remember seeing Jim Peebles in the car going up (he got the Nobel Prize of the year right before I did) on understanding the early universe. I remember worrying about all these solutions that mathematicians had come up with, different kinds of singularities, why cosmologists do not think about these other solutions? He looked at me and said: 'because the universe is not like that'. I thought my God, it is not, is it? I realised it is completely different and the question is why? I used to think that was something to do with quantum gravity. Why are they so different, why are the collapsed singularities so different from The Big Bang singularity? Big Bang is absolutely uniform (the gravitational degrees and the freedom are completely wiped out for some reason) and no explanation was coming out about that. I thought well maybe it is quantum gravity, like most people thought, but then it has to be a very, very peculiar kind of quantum gravity, which it is I suppose, but I could not really make that work until I was starting to use conformal maps, where big and small behave the same way.

Did you have a picture of an Escher? It describes a kind of geometry that is more or less irrelevant to what I want to say, it is hyperbolic geometry. [It] does not matter too much what kind of geometry it is, the point is the fish creatures. Look at their eyes, they are circles and they remain circles (they seem to get smaller and smaller) right up to the edge. Now that edge around the boundary represents the infinity of the world that these fish creatures inhabit. But the representation is what is called a conformal; it is due to a mathematician, Beltrami[12] (Poincarré[13] rediscovered it later). The Beltrami representation of this kind of geometry is very nicely represented by Escher. You can see by looking at the eyes of the fish creatures that they are circles; they remain circles close to the boundary. It is a kind of geometry where you can squash things down and big and small behave the same.

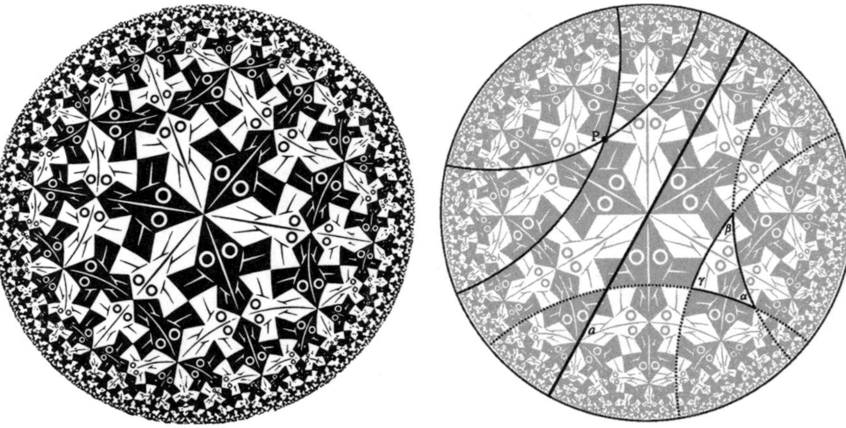

M.C. Escher, *Circle Limit I* and *Circle Limit I with lines*

You can do the same thing with physics so long as there is not any mass around, and this applies particularly to electromagnetism. The wonderful equations that Maxwell produced [which] described electricity and magnetism fit it all together and you can see light comes out of it; light is hidden in these wonderful, absolutely fascinating equations and they are conformally imbalanced. [For] big and small you can do one of these tricks, squashing, and the equations work just as well.

What makes the difference (when you can tell big from small) is when you have mass and this is combining the two most famous formulae of 20th Century physics. Einstein's $E = mc^2$ ('c' is just a conversion constant) tells you that 'E' (energy) equals 'm' (mass). Energy and mass are equivalent. Even earlier than Einstein was Max Planck;[14] he produced a formula $E = hv$ or hf. v or f is a frequency. Max Planck tells us that energy and frequency are equivalent. Put the two together, mass and frequency are equivalent. So, if you have a mass it is a clock. Every mass has a natural frequency attached to it – it wiggles very, very fast – it is much too fast for an ordinary clock so you have to gear it down. You gear it down and you have these wonderful clocks which now we have. It is how [you] make nuclear clocks which are extremely precise and they

12. Eugenio Beltrami (1835–1900) was an Italian mathematician best known for his work concerning differential geometry and mathematical physics.

13. Henri Poincarré (1854–1912) was a French mathematician, theoretical physicist, engineer, and philosopher of science.

14. Max Planck (1858–1947) was a German theoretical physicist whose discovery of energy quanta won him a Nobel Prize in physics in 1918.

come down to this fact that a mass is a clock. The converse of this is [that] if you do not have mass you do not have clocks, you do not have scales, you do not know big from small. [There are] two places in the universe where I claim you do not have clocks. One [is] at the Big Bang, because at the Big Bang energies get so huge that the mass becomes irrelevant, the energy in the particles moving around so fast is in their motion and the fact that they have mass is neither here or there, so you lose the scale. The other place [is] in the very, very remote future. I will not go into the details of that, but basically it is dominated by photons. Photons do not know the scale (they do not have any mass) and so in the remote future (so the argument

goes) again you lose the scale. [In] these two ends, the Big Bang and the remote future, the universe does not know how big it is. Then my idea was to stick one of these things (Big Bang to remote future) onto another one (another Big Bang remote future) and another one and that is my conformal cyclic cosmology, which I claim there is actually a fair amount of evidence for.

You can actually see through into things going on in the previous aeon, or what I call Cosmic Eon, [beginning] with our Big Bang. Certainly, the Big Bang was there, I am not complaining about that, although I was brought up on the steady-state model. Dennis Sciama used to try and persuade me quite convincingly that the Big Bang was a fake and it was not there. There was a good reason for it which came from a mistake in the observations. But the Big Bang was there and that is the beginning, end, and then what is the remote future? Well, it just expands and expands and gets less and less dense and becomes more and more boring and I will [begin] to think how boring this remote future is. Well, who is going to be around to be bored by it? Only photons[15] and it is very, very hard to bore a photon, not just because it probably does not have experiences but more importantly because it does not experience the passage of time. From creation to where it gets absorbed is zero time for a photon; that is just relativity. They go right to infinity and come out the other end (so that was the picture).

FÜ That was brilliant, thank you so much, Roger. I want to connect to what Carlo was saying about loop quantum gravity theory.

CR I thinking of that; when I said the universe bounces back, it is only a hypothesis. There is another possibility which is exactly what Roger described so far, that the universe loses [its] notion of size and then restarts again from a Big Bang.

FÜ Perhaps we could add a little bracket about your conception of time, because you say time is local and in a way in parallel to how Roger was saying a photon will be bored.

15. A photon is a tiny particle that comprises waves of electromagnetic radiation.

RP It has no time because it has no mass, so it does not experience any time... zippp... nothing.

FÜ So... is time real, Carlo, or is it a construct of our imagination?

CR No, time is not real; so, I am going to give you a five hour lecture about it. No, this is dangerous, taking sentences out of context. When Roger says that a photon does not experience time, it is saying something scientifically very precise which can be easily misunderstood. Not what we speculate about the world but what we know [to be] solid is what we call special relativity or general relativity. It is a vast modification of the way we know time works in the large universe, which has an effect that our usual notion of time is not good for talking about. It is of course not good for talking about the other possibilities, and so our common notion of time (which is the same for everybody, everybody becomes old at the same rate and so on and so forth), we know that is wrong. The time that we use for talking about the notion of time that we use in our daily life is not applicable either at the scale of the cosmos nor the scale of the small atoms and so on and so forth. This does not mean the time of our experience is not real, but if you look more in detail [about] how it works, it works completely differently from our usual way of thinking about it.

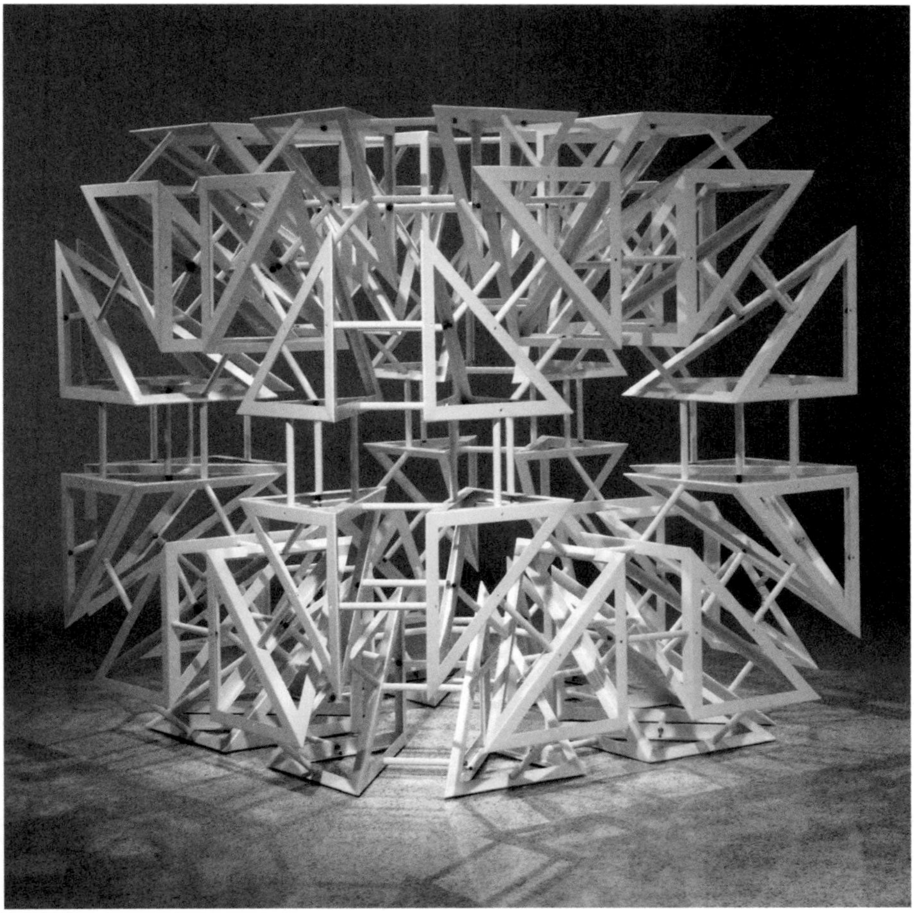

Lattices

Shawcross continued his investigation into the geometry of the tetrahedron - conceptually and historically the symbol of an indivisible unit of matter. Amplifying his interest in abridging different disciplines and forms of knowledge production and with a further inquiry on discovering a four-sided shape that did tessellate, Shawcross consulted a topological mathematician. He then created left and right handed tetroids, six of which formed a cube, and forty eight of which in turn formed these exploded cubic structures. These irregular, four-sided, structured shapes form a latticed system which in theory can expand forever. These expanding cubes are like Eadweard Muybridge's photographs, capturing an explosion or rapid expansion outwards from a single point. In this regard they are linked to the *Perimeter Studies* and *Plosions* in their attempt to envision the Big Bang.

[17]I. S. Shklovsky, *Cosmic Radio Waves* (Harvard University Press, Cambridge, Massachusetts, 1960), pp. 271-316.
[18]C. Hazard, M. B. Mackey, and A. J. Simmins, Nature 197, 1037 (1963).
[19]M. Schmidt, Nature 197, 1040 (1963).
[20]A. T. Moffet, Science 146, 764 (1964).
[21]M. Ryle and A. Sandage, Astrophys. J. 139, 419 (1964).
[22]I. S. Shklovsky, reference 17, pp. 369, 372.
[23]E. Fermi, Phys. Rev. 75, 1169 (1949).
[24]E. N. Parker, Phys. Rev. 109, 1328 (1958).
[25]T. H. Stix, Plasma Physics Laboratory Report No. MATT-239, Princeton University, 1964 (unpublished).
[26]R. W. Fredricks, F. L. Scarf, and W. Bernstein, to be published.
[27]V. Sarabhai, J. Geophys. Res. 68, 1555 (1963).
[28]E. N. Parker, Astrophys. J. 133, 1014 (1961); *Interplanetary Dynamical Processes* (John Wiley & Sons, Inc., New York, 1963).
[29]T. Gold, *Gas Dynamics of Cosmic Clouds*, edited by H. C. van de Hulst and J. M. Burgers (North-Holland Publishing Company, Amsterdam, 1955).
[30]E. N. Parker, Space Sci. Rev. 1, 62 (1962).
[31]M. Neugebauer and C. W. Snyder, Science 138, 1095 (1962).
[32]A. Bonetti, H. S. Bridge, A. J. Lazarus, B. Rossi, and F. Scherb, J. Geophys. Res. 69, 4017 (1963).
[33]E. J. Smith, L. Davis, P. J. Coleman, and C. P. Sonett, Science 139, 1099 (1962).
[34]N. F. Ness, C. S. Scearce, and J. B. Seek, J. Geophys. Res. 69, 3531 (1964).
[35]P. J. Coleman, L. Davis, and C. P. Sonett, Phys. Rev. Letters 5, 43 (1960).
[36]D. A. Bryant, T. L. Cline, U. D. Desai, and F. B. McDonald, J. Geophys. Res. 67, 4983 (1962).
[37]A similar increase in the low-energy proton intensity, in coincidence with the passage of a blast wave, was observed on 12 November 1960 [J. F. Steljes, H. Carmichael, and K. G. McCracken, J. Geophys. Res. 66, 1363 (1961)]. The increases proton intensity may have been partly or wholly of interplanetary origin, but the increase was of such a form that it is not possible to rule out a solar origin.

GRAVITATIONAL COLLAPSE AND SPACE-TIME SINGULARITIES

Roger Penrose
Department of Mathematics, Birkbeck College, London, England
(Received 18 December 1964)

The discovery of the quasistellar radio sources has stimulated renewed interest in the question of gravitational collapse. It has been suggested by some authors[1] that the enormous amounts of energy that these objects apparently emit may result from the collapse of a mass of the order of $(10^6\text{-}10^8)M_\odot$ to the neighborhood of its Schwarzschild radius, accompanied by a violent release of energy, possibly in the form of gravitational radiation. The detailed mathematical discussion of such situations is difficult since the full complexity of general relativity is required. Consequently, most *exact* calculations concerned with the implications of gravitational collapse have employed the simplifying assumption of spherical symmetry. Unfortunately, this precludes any detailed discussion of gravitational radiation—which requires at least a quadripole structure.

The general situation with regard to a spherically symmetrical body is well known.[2] For a sufficiently great mass, there is no final equilibrium state. When sufficient thermal energy has been radiated away, the body contracts and continues to contract until a *physical singularity* is encountered at $r=0$. As measured by local comoving observers, the body passes *within* its Schwarzschild radius $r=2m$. (The densities at which this happens need not be enormously high if the total mass is large enough.) To an outside observer the contraction to $r=2m$ appears to take an infinite time. Nevertheless, the existence of a singularity presents a serious problem for any complete discussion of the physics of the *interior* region.

The question has been raised as to whether this singularity is, in fact, simply a property of the high symmetry assumed. The matter collapses radially inwards to the *single point* at the center, so that a resulting space-time catastrophe there is perhaps not surprising. Could not the presence of perturbations which destroy the spherical symmetry alter the situation drastically? The recent rotating solution of Kerr[3] also possesses a physical singularity, but since a high degree of symmetry is still present (and the solution is algebraically special), it might again be argued that this is not representative of the general situation.[4] Collapse *without* assumptions of symmetry[5] will be discussed here.

Consider the time development of a Cauchy hypersurface C^3 representing an initial matter distribution. We may assume Einstein's field equations and suitable equations of state governing the matter. In fact, the only assumption made here about these equations of state will be the non-negative definiteness of Einstein's energy expression (with or without cosmological term). Suppose this matter distribution undergoes gravitational collapse in a way which, at first, qualitatively resembles the spherically symmetrical case. It will be shown that, after a certain critical condition has been fulfilled, deviations from spherical symmetry cannot prevent space-time singularities from arising. If, as seems justifiable, actual physical singularities in space-time are not to be permitted to occur, the conclusion would appear inescapable that inside such a collapsing object at least one of the following holds: (a) Negative local energy occurs.[6] (b) Einstein's equations are violated. (c) The space-time manifold is incomplete.[7] (d) The concept of space-time loses its meaning at very high curvatures—possibly because of quantum phenomena.[2] In fact (a), (b), (c), (d) are somewhat interrelated, the distinction being partly one of attitude of mind.

Before examining the asymmetrical case, consider a spherically symmetrical matter distribution of finite radius in C^3 which collapses symmetrically. The empty region surrounding the matter will, in this case, be a Schwarzschild field, and we can conveniently use the metric $ds^2 = -2drdv + dv^2(1-2m/r) -r^2(d\theta^2 + \sin^2\theta d\varphi^2)$, with an advanced time parameter v to describe it.[8] The situation is depicted in Fig. 1. Note that an exterior observer will always see matter outside $r = 2m$, the collapse through $r = 2m$ to the singularity at $r = 0$ being invisible to him.

After the matter has contracted within $r = 2m$, a spacelike sphere S^2 ($t = \text{const}$, $2m > r = \text{const}$) can be found in the empty region surrounding the matter. This sphere is an example of what will be called here a trapped surface—defined generally as a closed, spacelike, two-surface T^2 with the property that the two systems of null geodesics which meet T^2 orthogonally converge locally in future directions at T^2. Clearly trapped surfaces will still exist if the matter region has no sharp boundary or if spherical symmetry is dropped, provided that the deviations from the above situation are not too great. Indeed, the Kerr solutions with $m > a$ (angular momentum ma) all possess trapped surfaces, whereas those for which $m \leqslant a$ do not.[9] The argument will be to show that the existence of a trapped surface implies—irrespective of symmetry—that singularities necessarily develop.

The existence of a singularity can never be inferred, however, without an assumption such as completeness for the manifold under consideration. It will be necessary here, to suppose that the manifold M_+^4, which is the future time development of an initial Cauchy hypersurface C^3 (past boundary of the M_+^4 region), is in fact null complete into the future. The various assumptions are, more precisely, as follows: (i) M_+^4 is a nonsingular (+−−−) Riemannian manifold for which the null half-cones form two separate systems ("past" and "future"). (ii) Every null geodesic in M_+^4 can be extended into the future to arbitrarily large affine parameter values (null completeness). (iii) Every timelike or null geodesic in M_-^4 can be extended

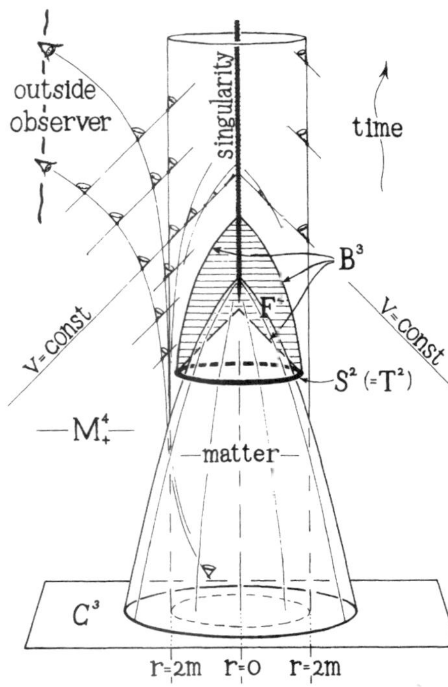

FIG. 1. Spherically symmetrical collapse (one space dimension surpressed). The diagram essentially also serves for the discussion of the asymmetrical case.

into the past until it meets C_3 (Cauchy hypersurface condition). (iv) At every point of $M_+{}^4$, all timelike vectors t^μ satisfy $(-R_{\mu\nu}+\frac{1}{2}Rg_{\mu\nu}-\lambda g_{\mu\nu})t^\mu t^\nu \geqslant 0$ (non-negativeness of local energy). (v) There exists a trapped surface T^2 in $M_+{}^4$. It will be shown here, in outline, that (i), $\cdots$, (v) are together inconsistent.

Let F^4 be the set of points in $M_+{}^4$ which can be connected to T^2 by a smooth timelike curve leading into the future from T^2. Let B^3 be the boundary of F^4. Local considerations show that B^3 is null where it is nonsingular, being generated by the null geodesic segments which meet T^2 orthogonally at a past endpoint and have a future endpoint if this is a singularity (on a caustic or crossing region) of B^3. Let l^μ (subject to $l^\mu{}_{;\nu}l^\nu=0$), ρ $(=-\frac{1}{2}l^\mu{}_{;\mu})$, and $|\sigma|$ $\{=[\frac{1}{2}l_{(\mu;\nu)}l^{\mu;\nu}-\frac{1}{4}(l^\mu{}_{;\mu})^2]^{1/2}\}$ be, respectively, a future-pointing tangent vector, the convergence, and the shear for these null geodesics,[10] and let A be a corresponding infinitesimal area of cross section of B^3. Then $[(A^{1/2})_{;\mu}l^\mu]_{;\nu}l^\nu$ $=-(A^{1/2}\rho)_{;\mu}l^\mu=-A^{1/2}(|\sigma|^2+\Phi)\leqslant 0$ where Φ $=-\frac{1}{2}R_{\mu\nu}l^\mu l^\nu$ [$\geqslant 0$ by (iv)]. Since T^2 is trapped, $\rho>0$ at T^2, whence A becomes zero at a finite affine distance to the future of T^2 on each null geodesic. Each geodesic thus encounters a caustic. Hence B^3 is compact (closed), being generated by a compact system of finite segments. We may approximate B^3 arbitrarily closely by a smooth, closed, spacelike hypersurface B^{3*}. Let K^4 denote the set of pairs (P,s) with $P\in B^{3*}$ and $0\leqslant s\leqslant 1$. Define a continuous map μ: $K^4\to M_+{}^4$ where, for fixed P, $\mu\{(P,s)\}$ is the past geodesic segment normal to B^{3*} at $P=\mu\{(P,1)\}$ and meeting C^3 [as it must, by (iii)] in the point $\mu\{(P,0)\}$. At each point Q of $\mu\{K^4\}$, we can define the degree $d(Q)$ of μ to be the number of points of K^4 which map to Q (correctly counted). Over any region not containing the image of a boundary point of K^4, $d(Q)$ will be constant. Near B^{3*}, μ is 1-1, so $d(Q)=1$. It follows that $d(Q)=1$ near C^3 also, whence the degree of the map $B^{3*}\to C^3$ induced by μ when $s=0$ must also be unity. The impossibility of this follows from the noncompactness of C^3.

Full details of this and other related results will be given elsewhere.

---

[1]F. Hoyle and W. A. Fowler, Monthly Notices Roy. Astron. Soc. 125, 169 (1963); F. Hoyle, W. A. Fowler, G. R. Burbidge, and E. M. Burbidge, Astrophys. J. 139, 909 (1964); W. A. Fowler, Rev. Mod. Phys. 36, 545 (1964); Ya. B. Zel'dovich and I. D. Novikov, Dokl. Akad. Nauk SSSR 155, 1033 (1964) [translation: Soviet Phys.—Doklady 9, 246 (1964)]; I. S. Shklovskiĭ and N. S. Kardashev, Dokl. Akad. Nauk SSSR 155, 1039 (1964) [translation: Soviet Phys.—Doklady 9, 252 (1964)]; Ya. B. Zel'dovich and M. A. Podurets, Dokl. Akad. Nauk SSSR 156, 57 (1964) [translation: Soviet Phys.—Doklady 9, 373 (1964)]. Also various articles in the Proceedings of the 1963 Dallas Conference on Gravitational Collapse (University of Chicago Press, Chicago, Illinois, 1964).

[2]J. R. Oppenheimer and H. Snyder, Phys. Rev. 56, 455 (1939). See also J. A. Wheeler, in Relativity, Groups and Topology, edited by C. deWitt and B. deWitt (Gordon and Breach Publishers, Inc., New York, 1964); and reference 1.

[3]R. P. Kerr, Phys. Rev. Letters 11, 237 (1963).

[4]See also E. M. Lifshitz and I. M. Khalatnikov, Advan. Phys. 12, 185 (1963).

[5]See also P. G. Bergmann, Phys. Rev. Letters 12, 139 (1964).

[6]The negative energy of a "C field" may be invoked to avoid singularities: F. Hoyle and J. V. Narlikar, Proc. Roy. Soc. (London) A278, 465 (1964). However, it is difficult to see how even the presence of negative energy could lead to an effective "bounce" if local causality is to be maintained.

[7]The "I'm all right, Jack" philosophy with regard to the singularities would be included under this heading!

[8]D. Finkelstein, Phys. Rev. 110, 965 (1959).

[9]The case $m<a$ is interesting in that here a singularity is "visible" to an outside observer. Whether or not "visible" singularities inevitably arise under appropriate circumstances is an intriguing question not covered by the present discussion.

[10]For the notation, etc., see E. Newman and R. Penrose, J. Math. Phys. 3, 566 (1962).

Excerpt from 'White Holes: Inside the Horizon'

Carlo Rovelli and Allen Lane

Think of a funnel. A very long funnel. At any given moment, the interior of the black hole can be imagined as being like this funnel.

The older the black hole, the more elongated its interior. The interior of a very old black hole might even be millions of light years in length. Here is an image of how we can think of the inside of a black hole at a given moment:

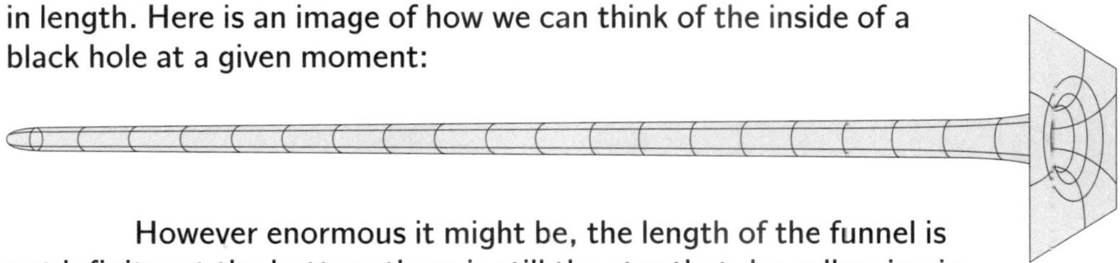

However enormous it might be, the length of the funnel is not infinite: at the bottom there is still the star that, by collapsing in on itself, gave rise to the hole:

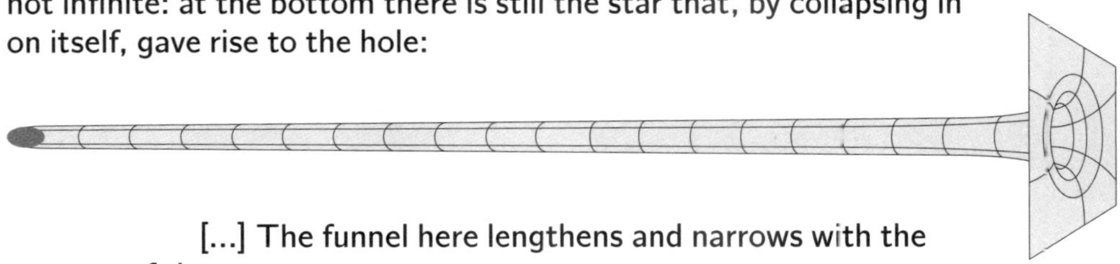

[...] The funnel here lengthens and narrows with the passage of time.

In order to illustrate this fact, the drawing below shows a succession of funnels representing the interior of the black hole at successive moments. Following the habit of physicists, the drawing shows a sequence in which time runs from the bottom to the top of the image. I do not know why this is the convention (perhaps we have borrowed it from geologists, who always draw the past below, because the oldest layers are deepest underground). The drawing must therefore be read from the bottom upwards: going up, the tube lengthens and narrows.

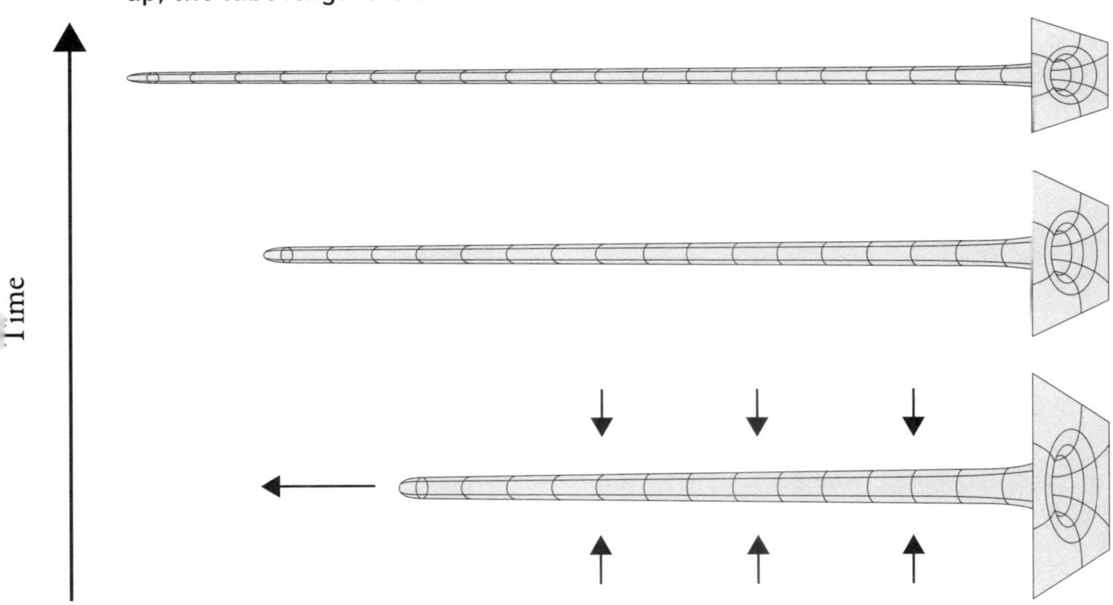

If we go down into the black hole, at each moment we are at a point in this funnel, going further and further down. Like this:

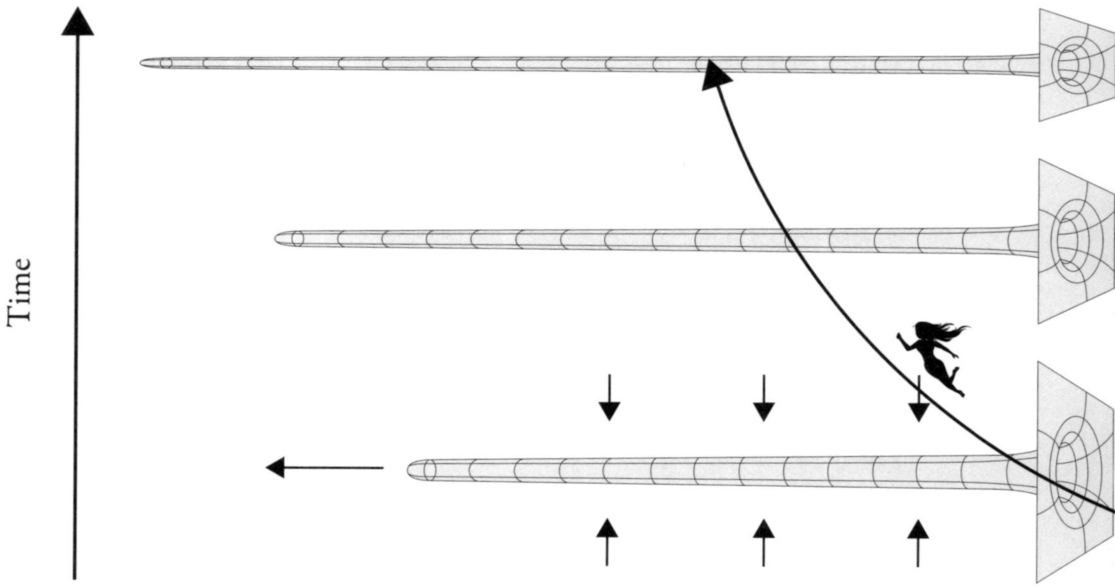

This is the shape of space inside a black hole: an endless abyss (dark and deep and nebulous it was)[1] that gradually constricts around us as we fall, without our being able to reach the bottom, where the star that gave rise to it in the first place has fallen.

How do we know all this, about the shape of the interior of a black hole, if no one has been there to see it and returned to tell the tale? We know it because the inside of a black hole is fully accounted for by Einstein's equations. Until something happens to create any doubt, we have no reason not to trust in these equations, since their predictions – spectacular and unexpected as they were – have so far all turned out to be correct.

[...]

It may seem natural to think that strange things happen at the bottom of the funnel, down in the centre of the black hole, in the dark area of the image:

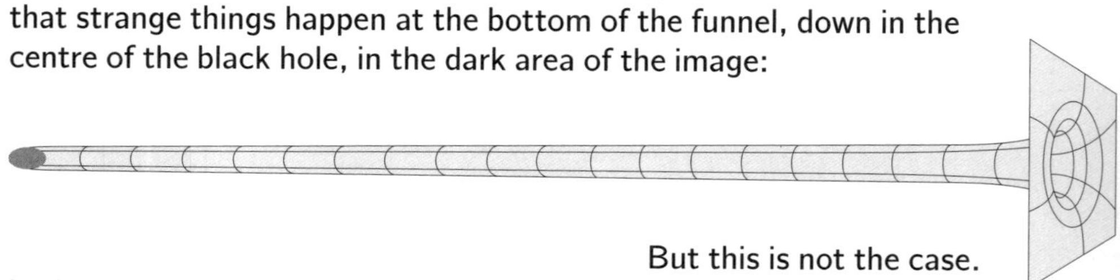

But this is not the case. In the centre of the funnel there is only the falling star; we are not in singularity territory. Here, the equations still work.

How so? If we go into a black hole that is very old, did the star not finish falling a good while ago? Has a long time not passed since it collapsed? A star that collapses on itself is crushed to a point in a

1. Dante's *Inferno*, IV.

very short lapse of time. So how come it is still there, still in the process of falling, after so long?

Time... It is always the crux of the matter. A 'long time' for one does not mean a 'long time' for another. A 'long time' for us does not mean a 'long time' for the star. Down there, at the bottom, time has slowed tremendously. Outside, millions of years may have passed, while down there just a few fractions of a second... The star is still falling at the bottom of the long funnel, which is stretching and narrowing, because in its time no more than fractions of a second have passed. So, the zone where the distortions become infinite, where the equations of Einstein stop working, the interesting zone... is not there!

So where is it?

It is in the future. It is in what happens after the interval of time depicted in the last image. It is in the dark grey area of this image:

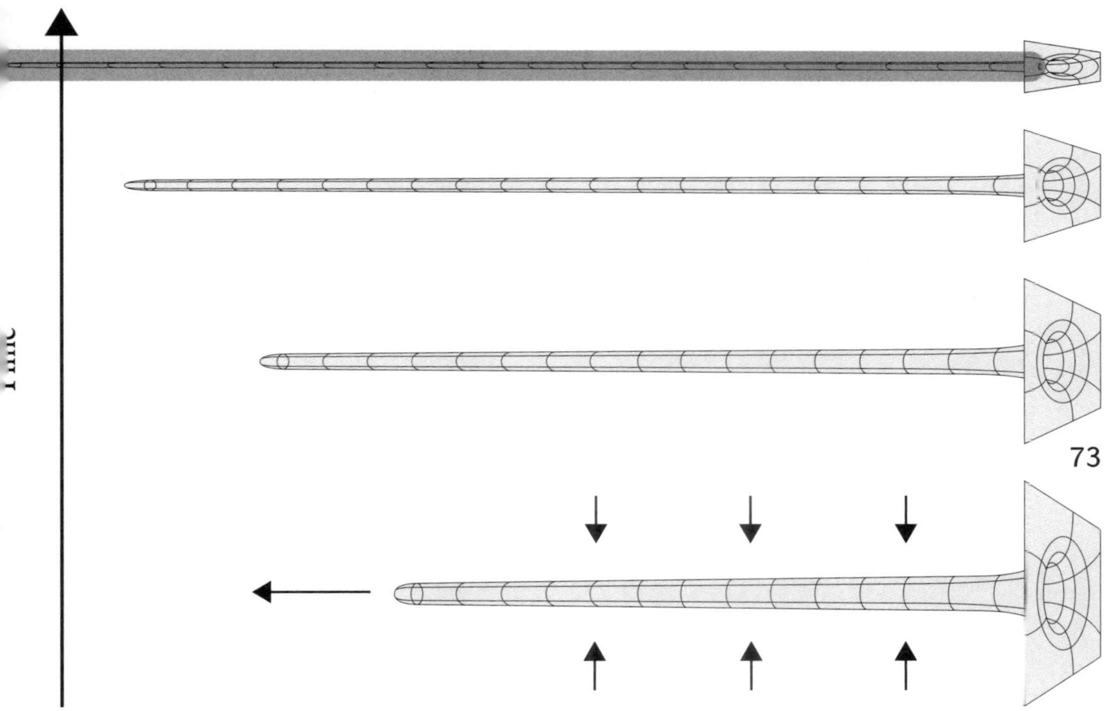

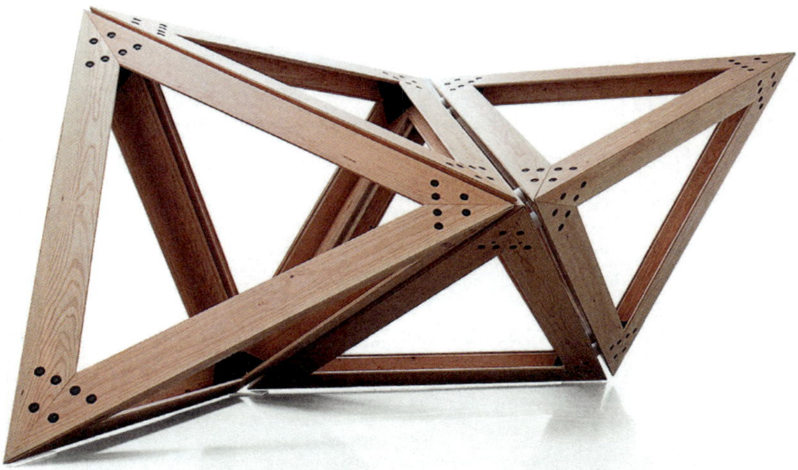

Axioms

Axiom was one of the last works that the artist created in his primary initial medium of wood. Conceived as an actualised epistemological metaphor, the work took as its foundation the tetrahedron, the simplest of the platonic solids, and the ancient Greek symbol of the atom. The *Off Cut Axiom* is an enlarged version of one of these building blocks, which together form a spiralling, lurching tower upwards.

Axioms formed the basis of the artist's development of *Paradigms* and, later, *Fractures*. The title *Off Cut* refers to the fact the work is formed from the remaining planks from *Axiom* (2007), a permanent commission in the Ministry of Justice in London.

Lecture Three, 28 September 2023: Patterns in Science and within Art

Speakers: Liliane Lijn, Marcus du Sautoy, Fatoş Üstek
Moderator: Conrad Shawcross

Conrad Shawcross

Thank you all for coming. As you know, this is the third of our lectures that are all part of this show that Fatoş has curated. Today, we are going to be talking about patterns in science and within art. As an artist, not a scientist, my intent often is to destabilise and create more questions than answers. As we have discussed previously, a scientific model will usually answer questions, but a successful artwork should perhaps raise more questions than it answers.

In this vein, my hope for today is [that] many people in the room who have entered thinking that they understand what patterns are, may leave with a different, altered concept of how ubiquitous, intrinsic, and fundamental they are. They are not just superficial visual entities, but invisible systemic forces that seem to pervade through every aspect of the world. It is deep within our human nature to search for patterns, the design contained within a system, sometimes when they are never there.

Now we are going to go straight into the questions, so Liliane will be first. Liliane has spent much of her career working on a particular form and this is quite in keeping with many ways that artists work. [The painter] Monet[1] returned to the same lily pads in the same pond at the same time of day, every day, and endlessly searched the permutations, the colours, the reflection of the light across the surface. [The conceptual artist] Sol LeWitt used the cube and endlessly explored the permutation of that, similarly to Carl Andre,[2] who employed the brick. [They] created endless geometric arrangements of those forms. I have spent 30 years working with the tetrahedron, which has been an endlessly rich vein for me, philosophically, geometrically, and conceptually. You can see many of those pieces within the show. This one is called *Off Cut [Axiom]* but Liliane similarly has worked tirelessly exploring one particular geometry which I think she will explain.

1. Oscar-Claude Monet (1840–1926) was a French painter and founder of impressionist painting who is seen as a key member of the modernist movement.

2. Carl Andre (1935–2024) was an American minimalist artist who is noted for his ordered linear and grid format sculptures.

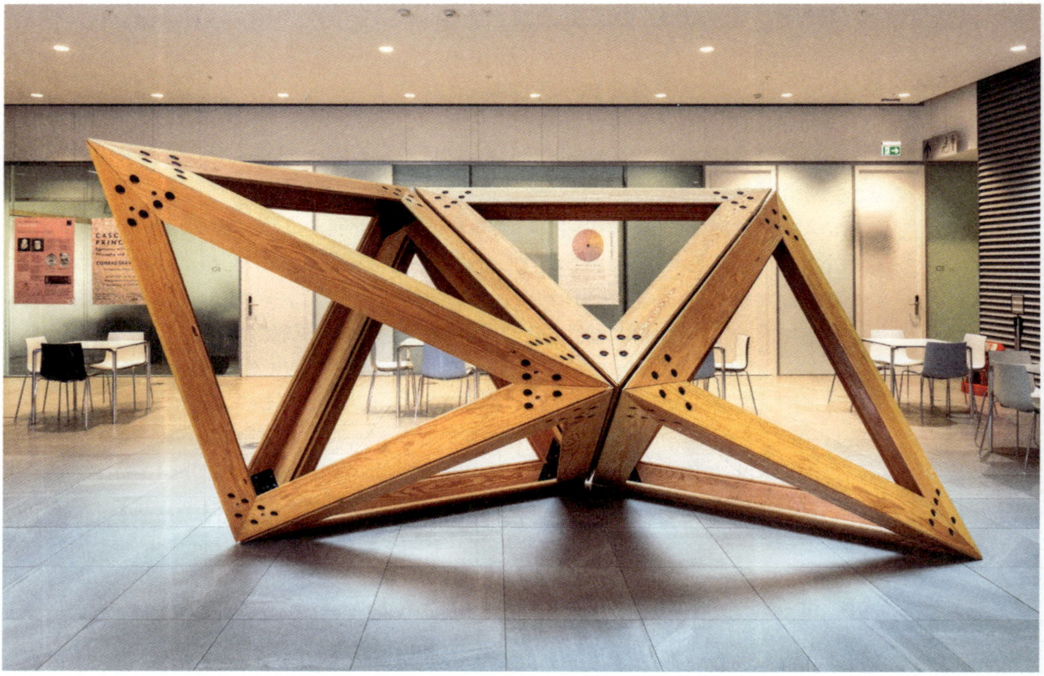

Off Cut Axiom, 2009

Liliane Lijn

In 1959, when I started working as an artist basically studying by myself, I was drawing and I drew a circle, and inside the circle I drew a triangle. The circle was full of strange creatures, the triangle was absolutely empty, and I thought, well, the circle really relates to my inner emotions and the triangles to my mind and I want to be that triangle and not that circle. I found that the relationship between the circle and the triangle is very interesting and I think for years I have been playing with that relationship or that pattern between the way the circle can transform into the triangle, the triangle into the circle.

I work with cones, but circles can become cylinders in three-dimensional balls and cones and spirals. This particular cone that you are seeing there (which is a video) is made out of fibreglass. It is about two and a half metres high, it is white, and it is dissected by planes of Perspex. The Perspex [is] inlaid, having taken out the material that [was] inside the cone, so we lay these layers, or planes of Perspex if you like, in elliptical tilted positions. What happens is very interesting. When the cone spins, there is a relationship, or you could say a pattern, set up between the different lines that are moving. In fact, the lines are not moving at all, it is the cone that is spinning and you are seeing different parts of the cone in motion. When you see it fluidly in motion it dissolves itself which is why I called it *Zero Gravity Koan*. Probably about 1972, I started making these aluminium and bronze solid cones that, when you touch them, transformed into organic shapes. What interested me was this capacity for a formal geometric shape to become an organic, living, moving shape and I thought that was another interesting pattern.

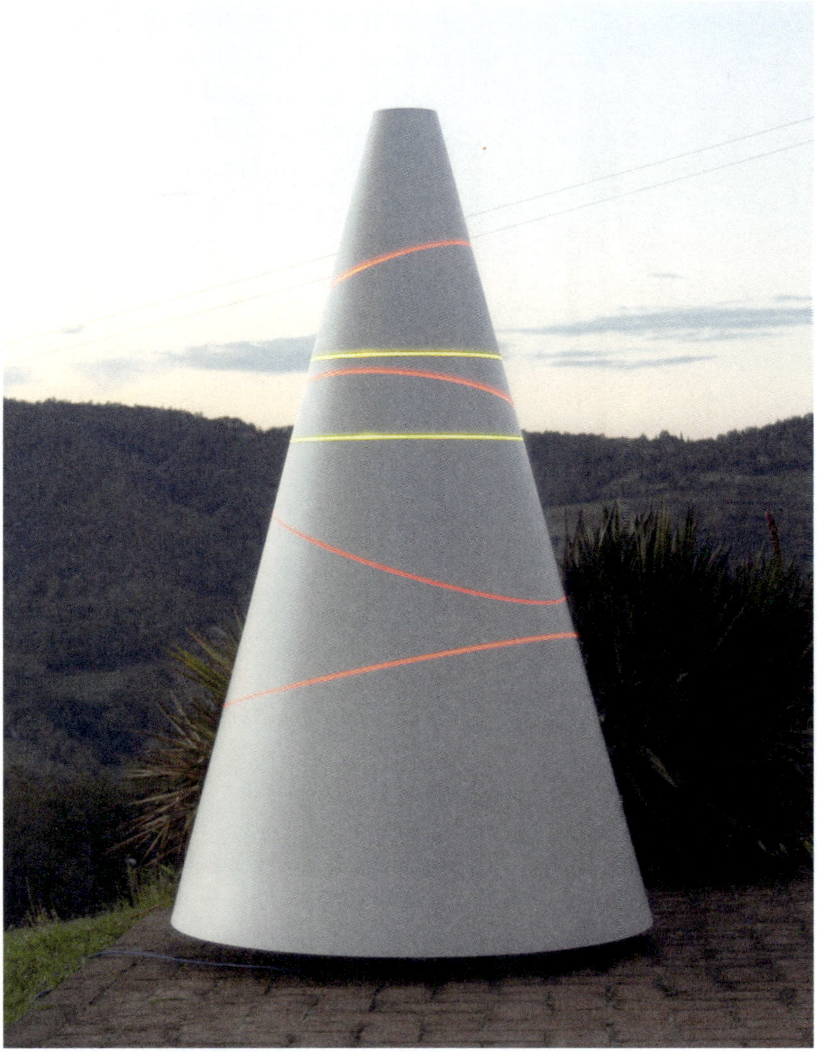

Liliane Lijn, *Zero Gravity Koan*, 2004

CS Similar to the tetrahedron, I think there is something so simple about the cone and yet so unbelievably complex. There are so many things that lie underneath the surface that these elliptical planes cut through. It keeps giving in terms of those amazing impacts and implications.

LL Well, cones are ubiquitous: there are the cones in your eyes that see colour, there are the traffic cones on the street, there are stupas, there are steeples, there are skirts, everywhere you see cones. I was told that the solar wind drops into the poles through conical forms and they call it the lost cone. That creates the Aurora Borealis, or so some scientists think, it is disputed.

This work that I have chosen to show you was made in I think 1982, and it is called *Split Spiral Spin*. In plan, it is a spiral made out of eight arms, eight parabolic curves, and it is made from perforated steel. It is 40 foot high.

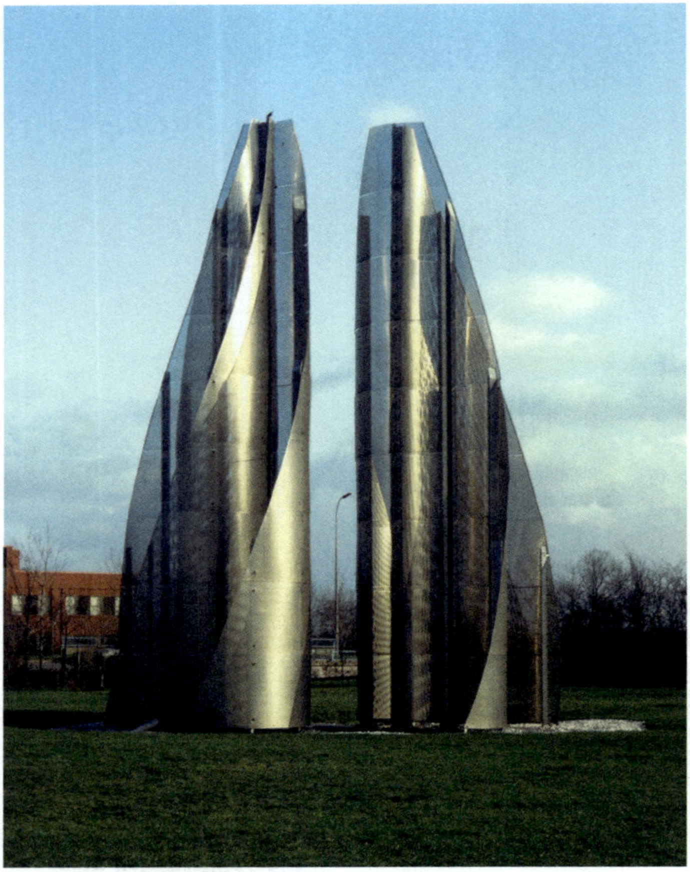

Liliane Lijn, *Split Spiral Spin*, 1982

CS You can enter into this. A sort of immersive experience, you walk through the moiré effect.

LL People in cars would stop and think they were looking at a glass structure, but in fact when the sun hits it, if you look at the other side it is opaque. So, you have an opaque and a transparent structure. The interference patterns are for me very interesting. I heard David Bohm talk about his theories at Birkbeck [University], and there were about eight physicists and me, I had been invited to go there. He spoke about reality as an interference system and that fascinated me and that was why I made this sculpture.

This sculpture which I made quite recently, in 2019, is a conical skirt; it is six metres diameter and made out of cloth. At its edge there is a strip of LEDs that goes right the way around the edge of the skirt. We spin the skirt but it is not simple, the spinning is not at one speed. You cannot accelerate it in a linear fashion. There you see it [video shown] beginning to spin. It is quite irregular and then the faster it goes the smoother its pattern. You can see the waves (there are actually three waves and there are always three) and if you look at it from above there are three spiral arms that whirl around with this piece, which I call *Gravity's Dance*.

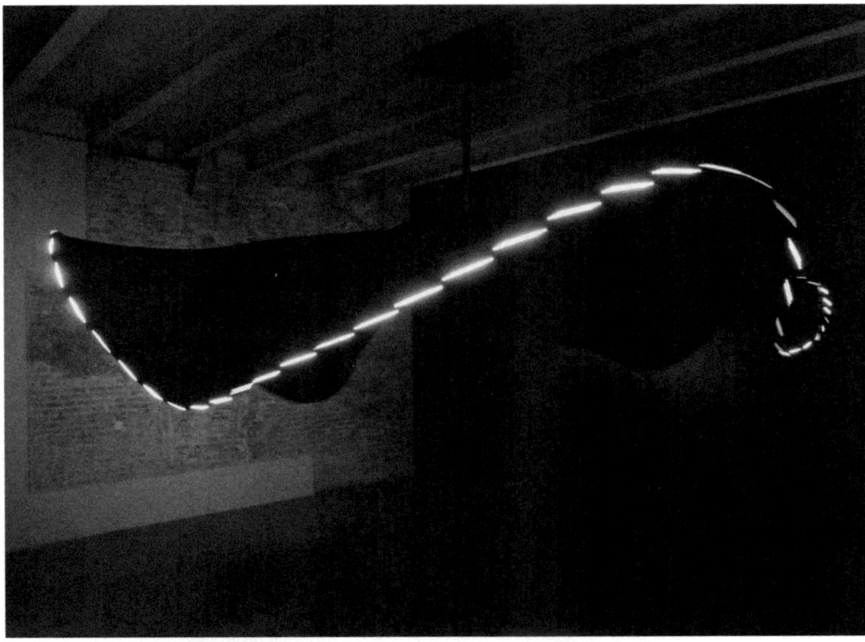

Liliane Lijn, *Gravity's Dance*, 2019

CS And is it hyperbolic? So, you have woven in more pieces than would lie flat?

LL [For] the first ones I made, which I called *Spinning Dolls*, I had four, one on top of each other. I also made it with a crocheted material. What was so interesting was that a crocheted material full of holes was actually much easier to raise than a solid material – when using a solid material it had to be heavy. It was extremely difficult to figure out the way the material reacted.

CS It does not seem to repeat, it is chaotic, wonderful.

LL It is quite chaotic.

CS When you lay it flat is it a circular piece of cloth?

LL It is a completely circular cloth.

CS You have not added one more slice to make it hyperbolic?

LL No.

CS It appears here like it is like a sea cucumber, that you have added one more slice of pizza.

LL It is just a circular skirt basically. My idea was the idea of a feminine skirt that becomes this almost diabolical movement, so in this particular piece it becomes quite dark.

CS It is like a whirling dervish.[3]

LL The whirling dervish are actually very light, very luminous, and they are people – there is no person here. I had not thought of the whirling dervishes, I had just spun the material and when I had so much difficulty and somebody said 'oh, the whirling dervishes', I thought well, what do they have that I do not have. I realised they have hips; that was the key. When I understood the hips then I understood that I had to have a flat, circular base to start with, clamp down the cloth on it, and from that it would rise, otherwise it went into a vortex. That is the curious thing about the piece which I showed in Pisa at the European Gravitational Observatory.

CS So, a key example of finding the solution in nature.

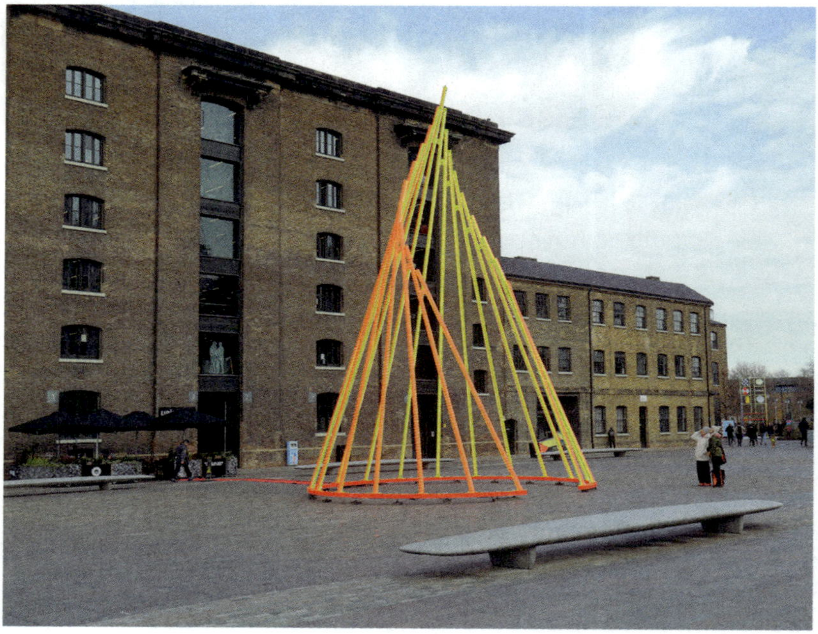

Liliane Lijn, *Temenos*, 2021

LL This piece is more recent again and it will be up in November [2023] again at Lewis Cubitt Square. It was commissioned as a kind of Christmas tree (I hate to say that) but I made this, which I do not think looks like a Christmas tree. It does light up at night and the reference to patterns is because I divide the circle [of the base] into 20 equal units and then opposite the centre I put a very tall pole which is 12 metres high, and then the next two poles; in increments I shorten the poles on either side, and they are attached to the central pole. Then the next ones are attached to the to the ones before and each pair are attached, so they are supporting each other. If you like,

3. Sufi Whirling is a form of physically active meditation which originated among certain Sufi groups. It is a meditation practice performed within the worship ceremony, through which dervishes aim to reach greater connection with Allah.

it is almost a self-supporting structure. The interesting thing is that the 20th pole, because there are 19 poles, the 20th pole, we had no place for it, so, it had to be left open.

CS It is a progression.

LL It is a progression that creates this arch, this kind of wing. Here you see it at night, it is coloured day and night, you actually see the colours during the day and then at night you see them again. And the black points are where the poles interfere with each other.

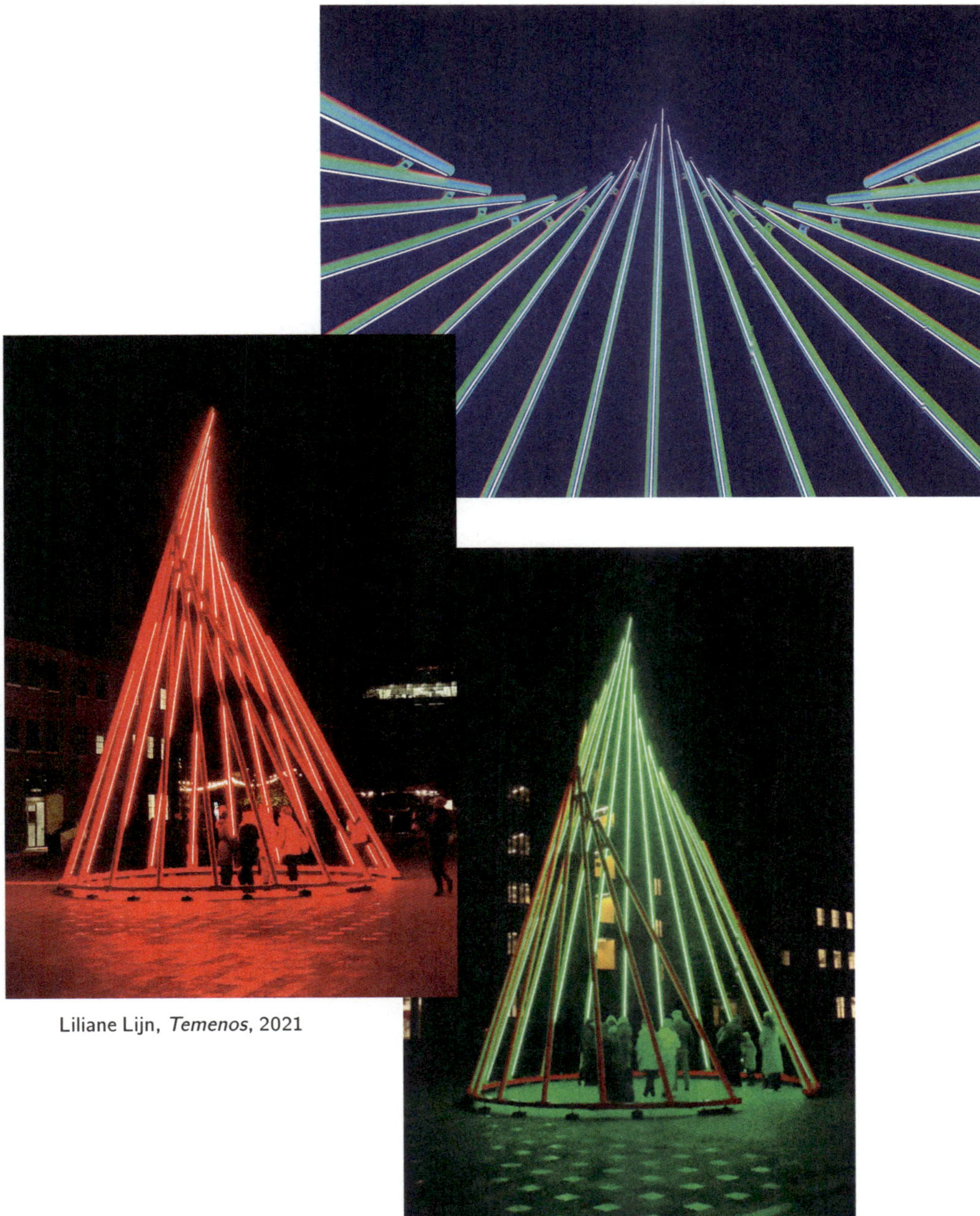

Liliane Lijn, *Temenos*, 2021

CS I have been doing a residency in Nottingham University studying the early universe, black holes, black hole ring down, and black hole mergers and looking at Lidar detectors[4] and gravitational waves. There is this almost universal representation of this journey as a sort of cone and so whenever I look at your works I really see this idea of expansion or duration represented through that thing. I think one of the great things about being in this wonderful context and with these budding, amazing mathematicians is that [with] these augmented minds you really can interpret them in such fresh ways. I have had so many endless interpretations of the work through the lens of mathematics here that I never would have thought of, but it is so valid in itself. These conical representations of space and time in your work for me, in my interpretation of them (which is just one interpretation), that was such an important thing with those.

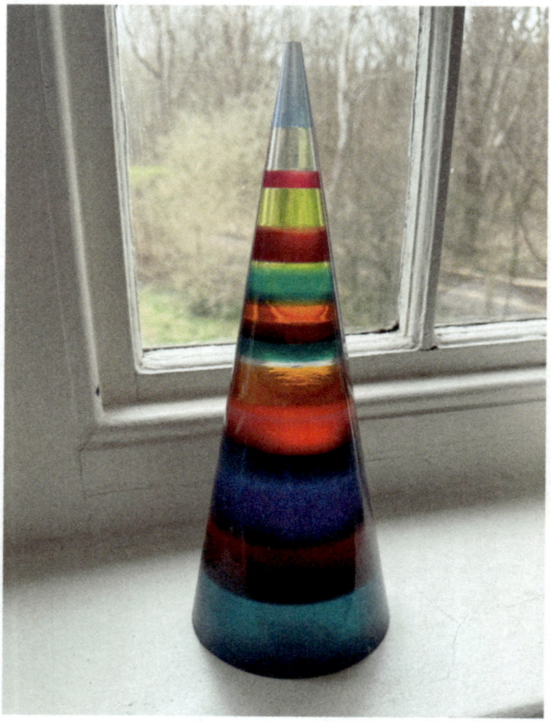

Liliane Lijn, *Liquid Koan*, 2003

I have one of your beautiful resin pieces on my windowsill in my bedroom, which I really love. It is also an example of how pattern is not just superficial, it is not just scrimshaw, but pattern is everywhere, it is not just a visual, two-dimensional, superficial thing, it is intrinsic. With those resin [cones] it is really lovely because it is opaque and translucent but you see the colours through every strata of the layers. It permeates through, it is not just a decoration, it is essential and fundamental.

4. Lidar is a method of determining ranges by targeting an object or a surface with a laser and measuring the time for the reflected light to return to the receiver.

LL In the late 1970s I wanted to get back into drawing which I had not done for a while, making the first cones that I made, and I started just by making little lines and I realised that they formed a pattern. What I was trying to do was just to use one specific shaped mark and that mark repeated, just by changing its angle, would completely change the way the drawing came out and so a pattern or a structure was actually created from this relationship between marks.

If you see the sculpture that I made much later, in 1988, that looked similar to DNA. Again, it is a perforated steel triangle, triangular V-shaped triangular forms, empty triangles if you like. Each one is shifted in very small increments and that shift is the only thing that differentiates them, otherwise they are exactly the same. This shift in their spatial arrangement creates this corkscrew spiral.

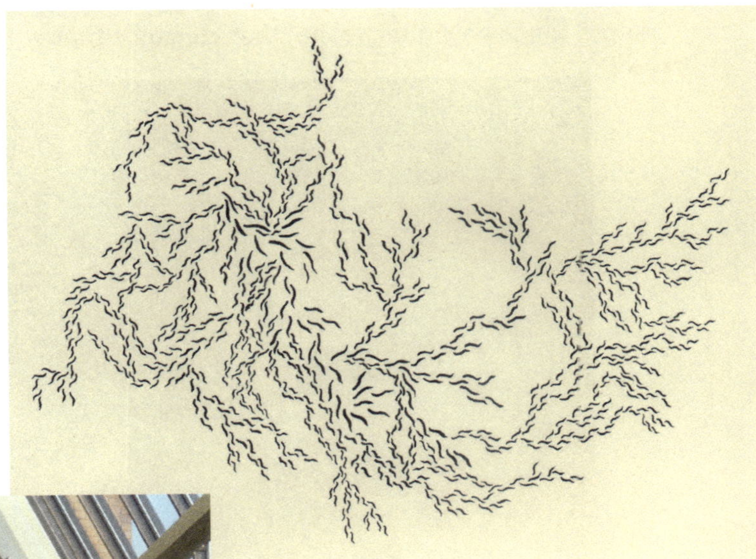

Liliane Lijn, *Flow Lines*, 1975

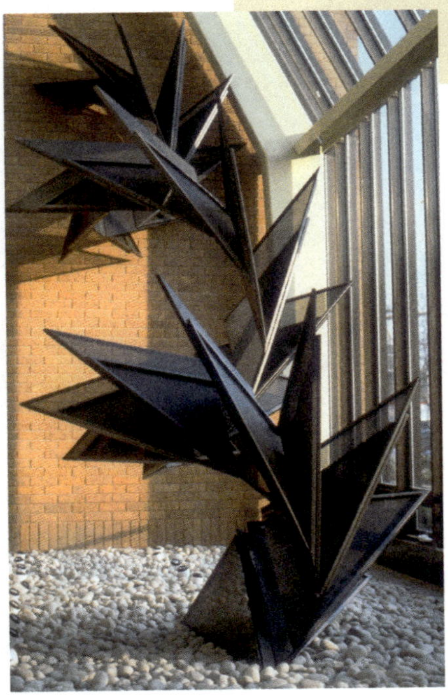

Liliane Lijn, *Carbon Black*, 1988

CS Now we are going to move on to Marcus, but just before we do I thought this would be a nice segue. This is a piece called *Loop System Quintet*, which is a piece I made back in 2007 and it is a series of five machines which take a very specific ratio. There is a 9:8, 5:4, 3:2, 4:5, so there is different harmonic ratios, and I use these as a springboard. The reason I put this on was when you sent me your slides I saw that beautiful toroidal light with the spinning dervish work (if I may call it that) and there was something very similar to this where there was this chaotic, but yet predictable loop. There is this sort of dichotomy with this because when you see these bulbs lurching around there is a sort of violent movement, but actually when you take a photograph there is something quite different. You suddenly see these steady, eternal, floating forms which are never changing, and these pieces create these beautiful toroidal knots which have that spline, that elegance that you had.

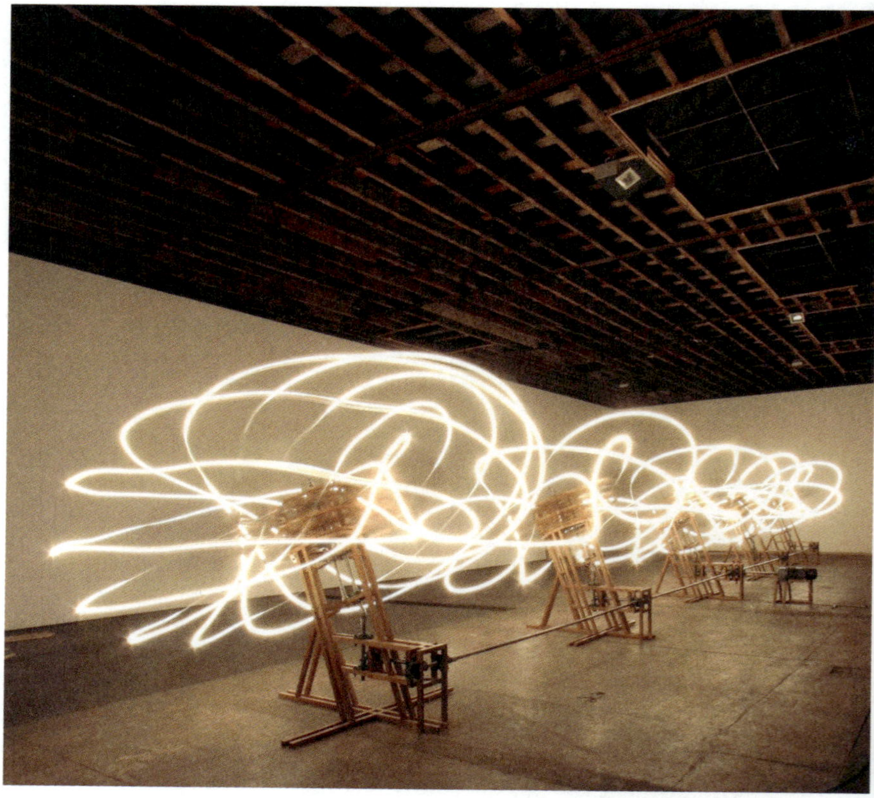

Loop System Quintet, 2005

This is a poor man's version of time splice, where I spent a very long evening, about 12 hours making a ten second loop where I filmed a piece, moved the camera, did a 30 second exposure, moved the camera one degree on a track, took another 30 second exposure, and took it again. By the end of the evening I was pretty dazed by watching this thing spinning around, but it creates this beautiful, eternal loop rotation. There is a spirit of investigation that is like a scientist's. This beautiful, floating, glorious thing came to me as a surprise; it was a real eureka moment. I felt like a scientist, like I have discovered something, but in fact I had not discovered anything. Maybe I came across something that already existed – that is [the] whole dilemma of invention and discovery.

I have this piece in Liverpool Street that is opening tomorrow, which is called *Manifold*,[5] and it is essentially a toroidal knot at the top that falls downwards, time is moving downwards. It is essentially a picture of a chord falling into silence. It was made of bronze in Spain a few years ago and it has been in storage for years but they finally finished the station. This is the culmination of those early researches with spinning a light bulb really fast and then taking a photograph of it, leading to these mathematical works.

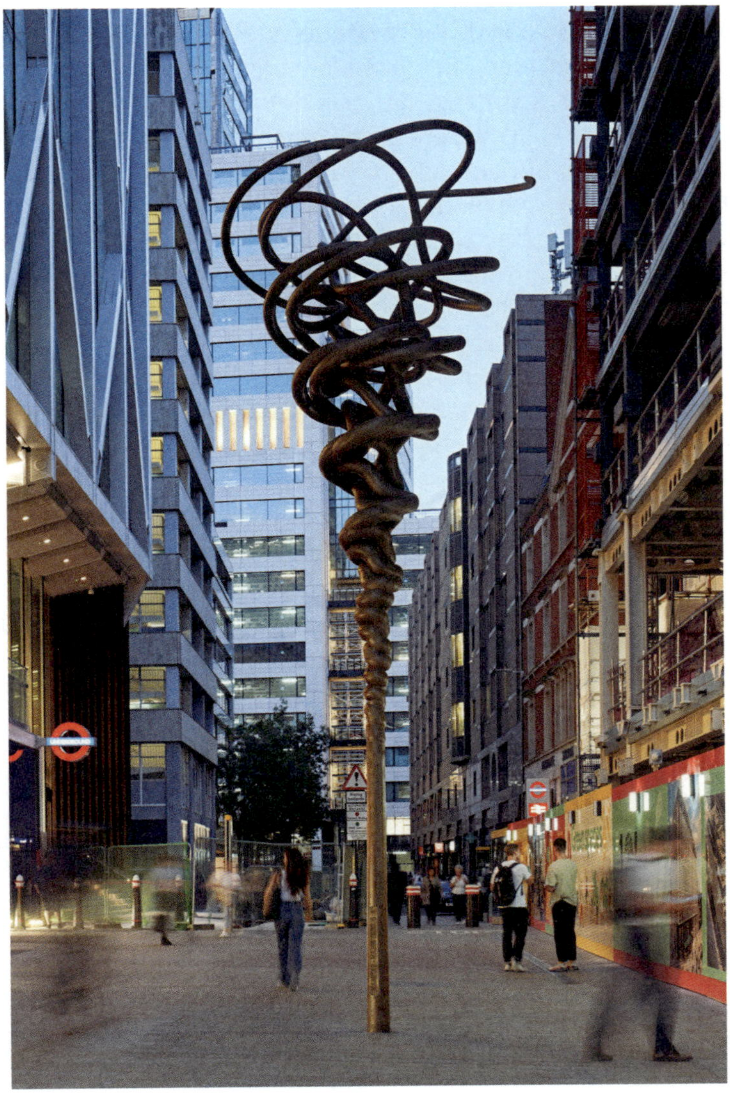

Manifold (Major Third) 5:4, 2023

5. *Manifold* is a sculpture in a series of works by Conrad Shawcross which reference chords in the harmonic spectrum.

Marcus, we are now on to you. I have been really enjoying reading your book again and I am so flattered to be mentioned in it. Your hero Gauss[6] has been on my mind a lot, the way everyone can interpret numbers in a different way. It is quite exciting to me, this idea that everyone in this room can think of a number, but see a different thing within it. That to me is something that I think maybe you can explain with these next slides.

Marcus du Sautoy

It was this word 'pattern' that intrigued me about our conversation because I think pattern for an artist immediately conjures up something visual, maybe two-dimensional, and for me it has a very different resonance. It is key to how I often define what I think a mathematician is, because I think people are who are not part of the mathematical world find what we do a little bit mysterious. Physicists bang atoms together in a large collider, chemists blow things up, biologists collect butterflies, but what is a mathematician doing all day in our offices? I think a lot of people think we are doing long division to a lot of decimal places or something, but actually I try to define a mathematician as a pattern searcher. I think that is what the essence of being a mathematician. Very often we are looking at the chaotic world around us, we are getting data from there and we are trying to see [if] there is any logic to this. Is there any algorithm, any pattern, rule that helps us to navigate what is going on?

So, I thought I would try and introduce you to the idea of patterns from a mathematician's point of view by giving you some numbers to look at and challenging you to see whether you can find the pattern. I am a big believer that mathematics is not a spectator sport, it is very important that you engage your mind. I have got a few challenges for you to get your brains into my mode of pattern thinking.

The first sequence: can you spot a pattern which will tell you what the next number is in the sequence 1 3 6 10 15 21 28?

Audience member 36

MDS Exactly, and how did you get the pattern there?

Audience member: It is quite easy.

MDS It is quite easy, I thought I would start you off nice and comfortably. This one you are just adding the numbers up one at a time. [There is a] very famous story of Gauss, who was able to spot a pattern which helped him to tell the 100th number in this sequence without having to add up all the numbers up to a 100, one of the shortcuts I talk about in my book. Maybe we will click on to the next

6. Carl Gauss (1777–1855) was a German mathematician, geodesist, and physicist.

slide because it tells us these are called the Triangular Numbers because there is a very visual way to think about it. A triangle can help you to find a formula because if you put two triangles together you get a rectangle and counting in a rectangle is easy. Times the width, the height, the base and so a formula would be half that number, we understand these numbers very well. The next sequence is one if you have read The Da Vinci Code you might have met. This one is a very famous sequence. What is the next number in this sequence: 1 1 2 3 5 8 13 21?

Audience Member 34

MDS Exactly, and you got that by adding the two previous numbers together so there is a pattern here. You build the next number out of the previous numbers. Your probably know a name for this, Fibonacci numbers. Actually, they should not be called the Fibonacci numbers because Fibonacci was not the first to discover these. He discovered them in the context of nature. In the natural world we find these numbers coming up all over the place because nature also builds things by taking what we have seen before and building on that, so they have a natural sense of growth. Which is why, if you take a flower and count the number of petals in that flower, invariably it is a number in the Fibonacci sequence. If it is not, that is because a petal has fallen off your flower, which is how we get round exceptions. Actually, these were discovered not by mathematicians and Fibonacci, but by musicians and poets in India who realized that there was a pattern to the way rhythm built up, if you use long and short beats. They wanted an algorithm to generate all of these complex rhythms of the Tabla[7] and they found that the Fibonacci numbers were the sequence that told them how they build. They should be called the Hemachandra[8] numbers actually. Ok, the next sequence: 1 2 4 8 16...

Fatoş Üstek 32

MDS 32 seems the most obvious answer, but actually 31 is the next number in this sequence. It looks like it is doubling but you need to know where your numbers are coming from. In this case, these numbers are actually describing something called the Circle Division numbers. If you take a circle and you put a point on that circle you just have one region but if I put two up I can join those up and I get two regions. If I add another point and join all the points up I have a little triangle and three regions around the outside, four. [If] you

7. The Tabla is a single-headed drum from the Indian subcontinent.

8. Acharya Hemachandra Suri (1088–1172) was noted for his expansions in fields like grammar, poetry, lexicography, and mathematics. Hemachandra described the Fibonacci sequence in 1150, 50 years before Fibonacci himself.

explore a little bit further, you add four points, you get this nice little envelope and four regions around the outside, eight. So, you think this is clearly doubling every time I put a point and join them up, probably everything gets divided, but then a surprise happens because when you have six points you can only squeeze a maximum of 31 regions out. For me this is important because mathematics would be very boring if everything was predictable.

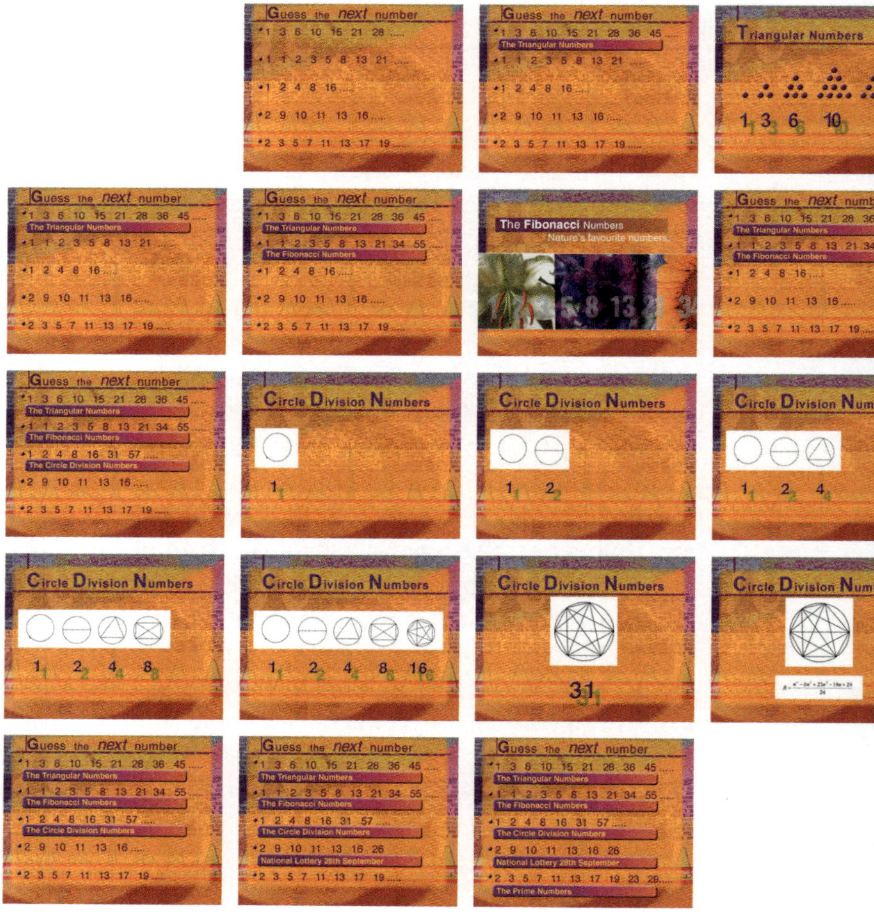

For me, the joy of mathematics are these sudden turns in direction, where a pattern you think is building up can suddenly go in a different direction. We have a formula for this but it is slightly more complicated than just doubling. You have to take the number of points and it is a quartic polynomial but we have understood the pattern and with that pattern we can make predictions. I think that is important because patterns are important to us as a species. They are our best tool for looking into the future, because if we understand a pattern, we can make a prediction about where we are going next.

How do we know about climate change? Because we have looked at the data and the data seems very clear about what we are facing. What about your diagrams of cones? How do we know that we have come from a singularity?[9] Because if we take the data back it collapses down to this singularity. Roger Penrose's last session was talking about how we can know what is going to happen in the future and, as he said, it is a very boring future where we just dissipate and just all become light. You are all getting very good at this.

LL That is the future, we become light?

MDS Yes, which actually is not too bad.

LL I wrote a book that said that [is] what will happen.

MDS You spot the pattern and there we are: everything is going to decay into pure energy gravitation. A little more challenging sequence for you... What about the pattern in this one: 2 9 10 11 13 16? You have all gone very quiet and I know there are some professors of mathematics in the room so you are probably holding back and saying: 'oh those are jolly easy; I will wait until the really difficult ones.' People are shaking their heads, which is good actually because these were in fact the lottery ticket number winnings for 28th of September last year. This is another important message. We saw that patterns can go in an interestingly new direction, but sometimes things do not have patterns, and that is a challenge. We so want patterns that we try and look for them everywhere. Fermat[10] and Pascal[11] produced the theory of probability to be able to tell things about even random [things].

My last sequence is probably one that you know about. It is the prime numbers.[12] These are the most important numbers in the whole of our subject, and these are numbers [where] we have been trying to look for a pattern to understand how they are laid out in the universe of numbers. We have been trying for 2,000 years and we still really do not understand how to predict where they are going to be. There are infinitely many of these scattered through the infinity of numbers. They get rarer and rarer, they never run out, but we are beginning to feel they are our greatest unsolved problem – one that people are working on here in the department, called the Riemann hypothesis, which is actually an articulation of the fact that these may not have an obvious pattern to them. That is actually quite difficult, to prove something does not have a pattern.

9. A singularity describes the centre of a black hole, a point of infinite density and gravity from which no object inside can ever escape.

10. Pierre de Fermat (1607–1665) was a French mathematician who is given credit for the early developments that led to infinitesimal calculus.

11. Blaise Pascal (1623–1662) was a French mathematician, physicist, inventor, philosopher, and Catholic writer.

12. A prime number is a number that can only be divided by itself and one without remainders.

So that is a little bit of, what for me, the word 'pattern' is about, understanding an underlying logic to something.

CS Fantastic, it is really fascinating. I very serendipitously had dinner with someone last night, I was telling them about this talk, and he gave me this amazing word that I had never heard of before and everyone I have talked to subsequently has not heard of it before. It is key to this lecture; there is this innate yearning of mankind to find the design or the pattern behind something. That relates to a sort of a creationist instinct or this idea that we spent 100s of years with Copernicus to try and find the pattern, platonic makeup, or architecture of the solar system, yet we finally realised that it was a series of catastrophic events and everything was quite arbitrary. We are still trying to take that further and find the secret of everything. 'Apophenia' means seeing patterns when there is no pattern and it is really interesting in regard to this lecture. It is really interesting in terms of those lottery tickets, or with the prime numbers, that the final result of all of this might be that there is no pattern, that the universe is potentially messy.

Like the *Schism* when you come in, no matter how many times you put those triangles together that form the 20 tetrahedrons, it will never quite reconcile, and you can try all you might but there will never actually be complete coherence, this grand design. Everything may be like the solar system, quite messy and coincidental. It is just a speculation.

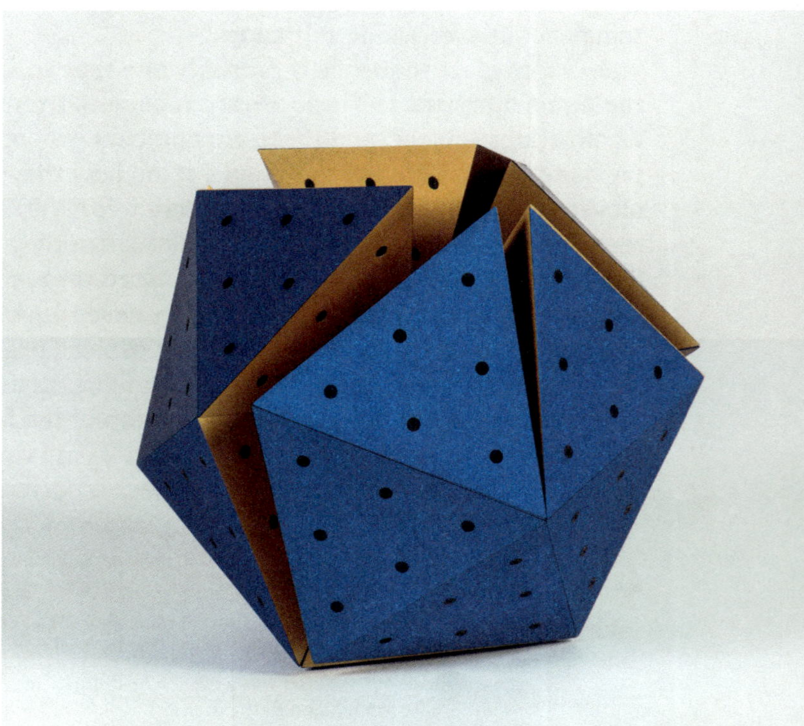

Schism (F24YB3), anodised aluminium, 2020

MDS I have a box upstairs which is full of letters from people with patterns that they think they have discovered in hugely varying areas. Very often they feel like numerology,[13] and scientists are generally very down on numerology because that is generally associated with the desire to find patterns where there are not, but actually I am less down on the numerologist because I think at its heart is the spirit of what we do. We are open to the idea that maybe something bizarre is happening here.

Kepler came up with this mad idea that the way the planets were arranged in our solar system had something to do with the way the platonic solids could be put inside each other. It looked brilliant, it perfectly explained the number of planets we could see, the distances between the solids, and he even built this with all of the platonic solids, one inside another. Then of course we discover several more planets and the platonic solids have run out. We have only got five and the whole thing is blown out of the water. But I think that spirit of just trying to look for an explanation, a pattern to explain how the planets were, of course that is actually the same spirit that led to him understanding the laws that control the way the planets move, Kepler's laws.

LL Do we follow patterns unconsciously in any case? Our brain is constantly (even though unconsciously) looking for a pattern. We have to function and our functioning is through this automatic visualisation of patterns without actually seeing them.

For example, one day, going out to dinner years ago I remember having a little bit too much to drink and then driving across Hyde Park Corner when there were no traffic lights. I just remember driving and thinking: 'My God, how am I doing this?'. You know, cars coming in every direction and yet we do it very easily.

MDS The other cars may have had a very different view. I think the brain has evolved to be mathematical at heart, this element of looking for patterns is how we have survived. Those that did not spot the pattern were unable to look into the future and act accordingly. There is a big argument about consciousness and the fact that a lot of the illusions that happen to the brain are because we have to fill in so much, due to the patterns we have seen in the past. If something breaks that pattern it has to reassess itself in relationship to the outside. So, it is how we have to operate.

You talked about time quite a lot with Carlo and Roger in the last session and Roger was talking about the fact that photons do not have a concept of time, but how do we have a concept of time? I think that is highly related to the idea of a pattern because something needs to repeat itself in order [for] you be able to understand a unit of something you are going to compare everything else against.

13. Numerology is the belief in an occult, divine, or mystical relationship between a number and one or more coinciding events.

LL Young children have no concept of time. It takes quite a long time before a child actually understands time.

MDS Because they have got to build up that recognition of something repeating. There is evidence that the caves in Lascaux[14] are actually a calendar where the dots that you see on the wall (obviously there are wonderful pictures of animals and things), but the dots on the wall are actually about repetition of when the stags are rutting or where they are mating. You have got the Pleiades[15] right at the beginning of this sequence and it is saying, look, these are repetitions in the calendar and therefore this is how we can have a concept of time.

CS Speaking of time, Fatoş you did this very intense project which [was] an incredible show that was 50 exhibitions in 50 weeks. So, it is very much about indexing in a pressured time situation.

FÜ To be honest with you, there are so many things I would like to respond to, and it is very nice to be in the speaker's seat today.

Ok let me focus on my presentation. I think I was very obsessed with time and the notion of being in the now. That obsession led me to do a trilogy of exhibitions. I even ran an art magazine that was called *Now is Were*. It is not like now is 'where', it was actually whenever you say 'now' it is singular, but whenever you mark that 'now' with some associative experiences, it becomes a plural past. All of our writers of course misspelled it, but it was more of a conceptual exploration of what is happening in the visual arts. I think for me what has been very interesting is studying mathematics and recognising patterns, but at the same time it is the euphoria of finding one.

I think I will first talk to you about this project which is unprecedented artistic research that I did 11 years ago. I used to teach at Linkopings University in Sweden and they had this department called Vision Forum which was a really open department. I led a couple of projects that were artistic research projects and they were open ended. We did not need to come up with an exhibition or any outcome and we could just do whatever we wanted. With a curator colleague, Anna Gritz, we actually started *Infra*. If you know the artist Marcel Duchamp, he is one of the important fathers or thinkers of contemporary art. Marcel Duchamp came up with the concept of 'inframince' (infra-thin). That could be the dust that collects in your shirt pockets or the warmth of the bus seat. It is this ultra-thin layer that is connecting different realities. As you see here, we invited six artists with Anna and we wanted to treat ourselves like empty vessels so we sat in an economics class on matchmaking

14. Lascaux is a network of caves near the village of Montignac, southwest France where early cave art was famously found.

15. The Pleiades is one of the most easily recognisable star clusters in the night sky, a constellation named after the seven daughters of the Titan, Atlas.

theory at Cambridge University. We sat in an astrophysics class at Imperial College, but then we also had training with a Gurdjieff practitioner. Gurdjieff made this anagram and he says there are nine patterns in humans' lives and we are born with those patterns; it is what we are constantly evolving around. It is very esoteric but there are special movements in your body that you could do, to overcome what is limiting you. We also even played the game Go among many other activities. I have quite a fun one-minute video which was made by one of the artists who was participating in the 'Infra'.

[Voiceover: We were invited to take part in a project called 'Infra'. The plan was to sit in on established formats for knowledge distribution and activities. Everything would be connected; otherwise, the details of the process of the coming year was a complete mystery.]

It was a one-year research project where we were trying to find what connects different formations of knowledge production. The artists were responding to the fact that we started with finding Masonic traces in London. Then it was really stretching the content as much as possible from the spiritual to the hardcore scientific.

I have been mostly working with different temporalities of curating and this project really won me the badge of crazy. I curated 50 exhibitions in 50 weeks with an 11 hour turnover from one show to another. It was at the ICA in 2015. I had this one room, but at certain times of the year I could also use the ICA theatre, and then also I commissioned artworks at Bicester village, which is just around the corner from here and received 60,000,000 visitors before covid (second best visited place in the UK). So, this, *fig-2*, is basically a handed over brief to me from what happened 15 years ago. In the beginning, the year 2000, *fig-1* was curated by Mark Francis and that became an epic show in the sense that it really created the buzz that the London art scene needed, before the term 'pop-up show' was coined. At the same time, it was really giving a platform to young contemporary artists because most of the gallery scene was still looking to Impressionist paintings and other classical modern art forms. With *fig-2* it was really about thinking how could I create a platform for artists to be as experimental as possible and really create a non-institutional platform for them to take risks.

I mapped my curatorial methodology so we could perhaps go there because I can talk about *fig-2* for five hours or five days depending how much time you have. It was quite an epic project; we had 50 exhibitions and 148 collateral events. *fig-1* was one exhibition, one artwork at a time, that was taken out from a studio [and] brought into the Georgian townhouse that it was using. For me in 2015 just doing the replication of *fig-1* was too boring so I thought I have 50 dots, as you see on this beautiful paper. I was interested in how I can create a continuity, so I thought I could actually have an overarching theme that connects all the exhibitions and then I could perhaps have a linear connection [to] what the previous exhibition did. The next exhibition could in a way respond to it, expand it, or take it further, or sometimes introduce a sharp U-turn. First of all, I was so ambitious that I thought all my family and friends would come [to] all my shows (yet, they did not), but there were some people that did. We had a loyalty card that we would stamp every week. When I moved to the UK, I was so perplexed about loyalty cards and probably for the first five years I lost a lot of money and free coffees just because I never collected one. So, I thought we will have a loyalty card and they got a free publication [when this was filled up].

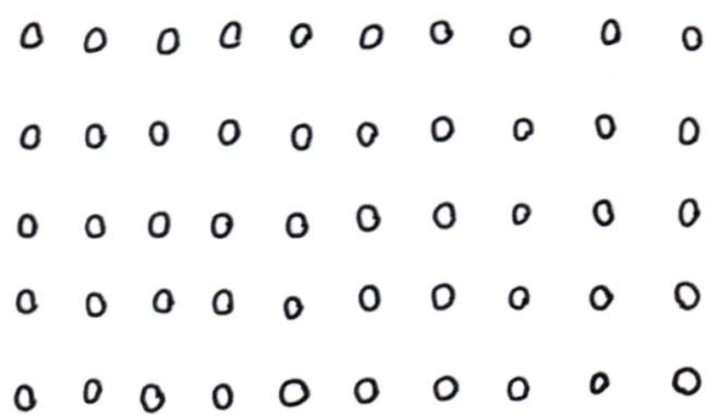

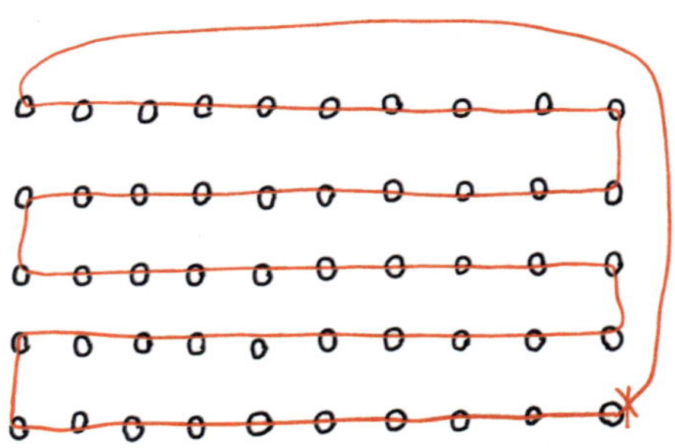

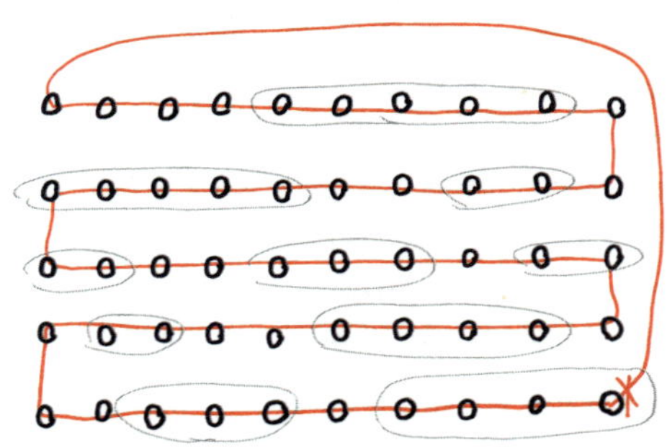

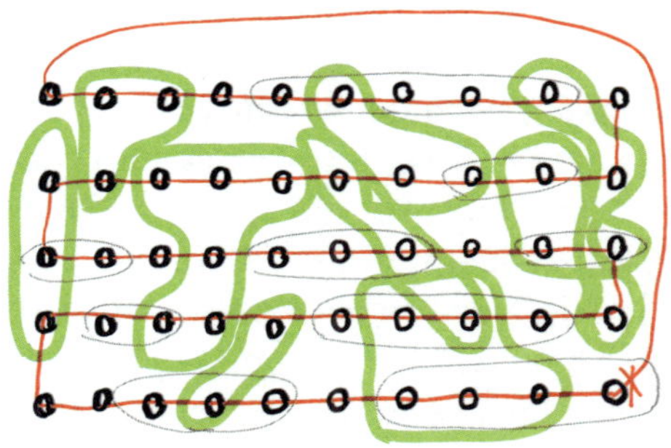

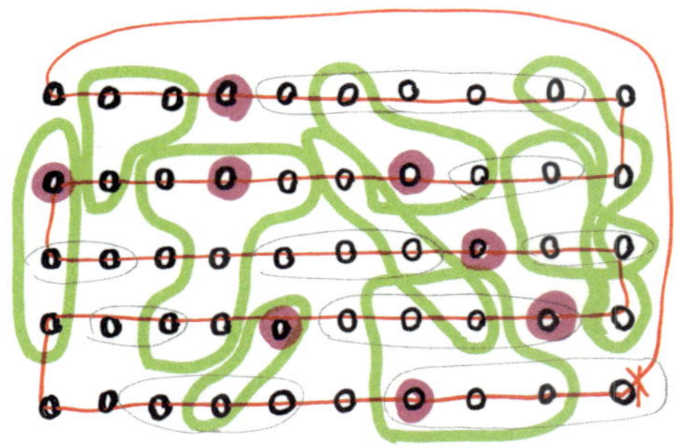

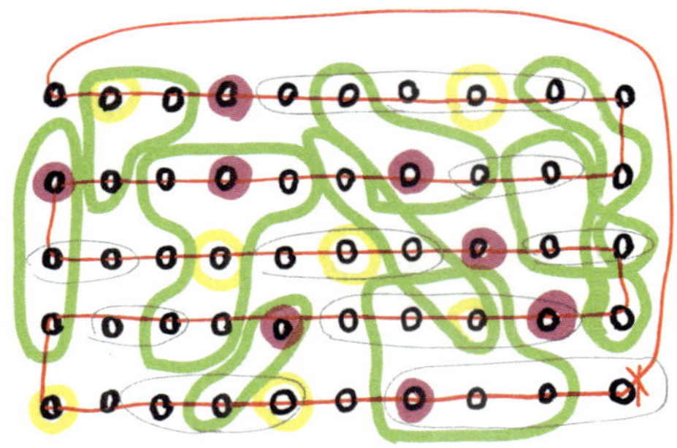

I thought I am going to connect all the exhibitions and borrow a method from what *fig-1* did, which started with a solo show by Richard Hamilton, showcasing an in-progress work, and Hamilton bookended *fig-1* with the completed painting. I thought that was a brilliant idea, so I started with someone who is young and who had never had a solo exhibition in London before. 'I will show an in-progress work with them and then end the exhibition with it.' Then I thought that is not enough. I studied topology while I was an undergrad maths student so for me linearity is interesting but then it is predictable. Then I thought maybe it would be great if I could have these vague groupings; I started with 'what is in a line' and that line in the first one was made into carpets. The second one actually was by Charles Avery so it was in drawings. He is very much interested in non-Euclidian spaces. The third one was actually a representation of memory, the fourth one became a prose and poem etc. Then I thought okay I could have areas where I have a focus on a fashion in art discourse or performance as a medium or different things, so I started doing these groupings.

I thought that was not enough. So, you know if you have played the game Snakes and Ladders, sometimes you throw the dice and you go to number 42 but then you fall into 17. I thought not everyone is going to see the whole 50-show offer which was very important. Then how could I still create a sense of an ethos of a project where you came only to week two, week 47, and week 28, and still had an idea of what the whole project was about. I started creating connections that were almost nonlinear collections between weeks. Let us say if the artist of the third week was talking about line as memory then I would bring another person at week 37 that responds to some of the raised concepts... But that was still not enough. I thought, okay, I need to punctuate something so that there is this confluence of artistic practices coming into one another. Sometimes I would also invite scientists to collaborate with artists or sometimes it would be cybernetic experts or people from all these different areas. I thought I need cluster points so that there are almost gravitational areas where people feel there is something more about it. Then I wanted to bring a discursive expansion to the project. That was my curatorial seminar programme. It was eight seminars that were all about looking at curatorial practice. We also did workshops for UK based directors and curators.

This was actually my curatorial methodology for the 50 exhibitions and we also have a diagrammic representation of it. I thought this would be good to showcase my patterns of thinking.

MDS Why did you chose 50?

FÜ I did not. 50 chose me.

CS It is so interesting to hear you talk about this after I saw these slides, these 50 dots. They are like a wonderful 1950s drawing of an amino acid. There is something very logical about it but there is something also a bit crazy about it which I love. If I was a doctor specialising in apophenia and laid out all these dots and ask people to draw, I would say you have a bad case of apophenia.

MDS It reminds me a little bit of George Perec's *Life: A User's Manual* because of the idea that he has a 10 × 10 apartment block and he puts these rules on (he wants he makes a Latin Greco Square out of it) so that each room has to have one thing from this thing one from this thing and that constraint leads him to incredible creativity. Then the way you go around is dictated by a knight's move and then there is one room missing. I think those kind of constraints are what is really exciting. You force a pattern on it and it can take you somewhere really interesting creatively.

FÜ George Perec is really an influence because he was part of this OuLiPo movement, this French literature movement, the *oeuvre of potential literature*, if we could bluntly translate. Like Georges Perec they would all [give] themselves obstructions and then those obstructions would be the ground for their wild creativity. George Perec had a writer's residency at MIT Computing Labs. He was invited to write a novel based on binary code; the book is titled *How to Ask for a Raise from your Head of Department*.

MDS Maybe I should be reading that.

FÜ It is such a good one, it is hilarious, and written in a binary code. He also wrote this novel in French without using the letter 'e', which is the most frequently used letter in French and then there is a fantastic translation to English – the whole book is written without a letter 'a'.

CS As artists, as designers, sometimes these constraints are very liberating. It is the liberty of constraint. Sometimes design needs constraints in order to progress.

I was going to show you this object as an end of the lecture. This piece has emerged from 20 years of trying to develop interference patterns and disrupt surfaces like you, with these moiré patterns and optics, letting the sun transform surface as a form of camouflage.

I have this piece called the *Patterns of Absence* and this is all about interfering holes. There are 100s of 1000s of holes and these two discs are rotating very slowly.

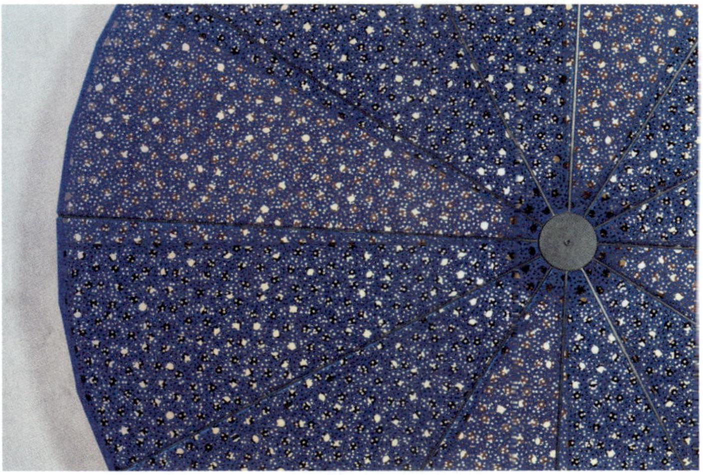

Patterns of Absence "M" (Bm15D7), 2021

This box is called the *Library of Absence* and it is inspired by a 1950s Dansette that my dad gave me when I was a teenager. It still works; it has got wonderful valve technology in it, and I still have it in my house and play 45s on it. I have got this skin which is quite reminiscent of that and the whole thing is a briefcase. You can carry it around, it is anodised aluminium and then you open the box up, it reveals a contraption inside, and then here we have 64 interference patterns. These are all unique patterns I have collected over the last 20 years. They are all different and they are sort of colour-coded. Essentially, you have got a light and it counter-rotates. It achieves what you are seeing up there but we have got these endless different permutations. The classic golden rule with moiré is that the patterns have to be the same for them to create the holographic effect. If you change the scale of the pattern, the effect is destroyed. You can place them further apart, you can change the pitch, but they have to be the same for them to interfere with each other. It is quite fractal in that way; if you do not feed the same thing back into itself it does not work.

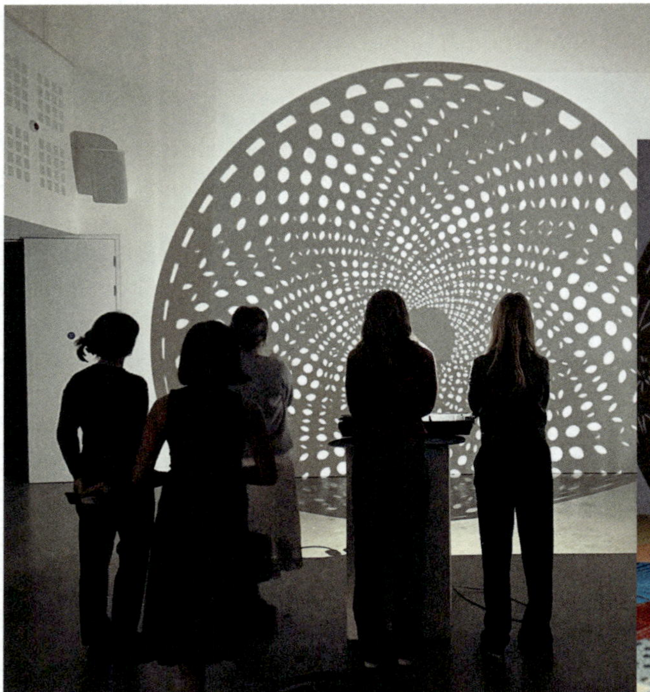

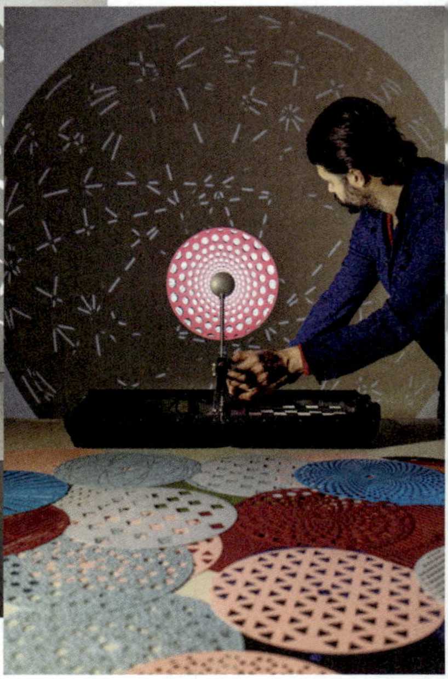

Library of Absence (Éditions Take 5), 2023

You could turn the lights off now. So essentially, it is a sort of library of all these different perforations. If I then try a different one then the thing that becomes unpredictable is you can mix discs together which then creates a very beautiful interference. Some of the interesting ones are the simplest ones in repetition. I think there is a mathematician called [Stephen] Wolfram[16] who deals with this these ideas of interference and complexity. Apparently, his argument is that the simpler the patterns in repetition, the more quickly you arrive at complexity. The more complex the patterns are the longer it takes for you to reach [complexity]. These are just simple slits. It is very surprising how much flux and change happens within these patterns and it moves in different directions, which again is very surprising. This is the one that I regarded as probably the least interesting before I tried it but yet it has almost the most complexity of all of them because it is the simplest.

The other thing is that you can play with these colour discs and you change the colours and you can mix them together so this becomes an optic thing in itself. You get the dilution of reds and greens and blues through the disc so you get the same interference you would get on the wall-based piece.

This [*Library of Absence*] is a piece that is going to be produced by a Swiss bookmaker called

16. Stephen Wolfram (b. 1959) is a British-American computer scientist and physicist, who began his professional life at the age of 15, researching particle physics and applied quantum field theory.

Éditions Take5. This is the first time we are showing it. It is going to have a book in here and the discs in there. It is the most sophisticated design we have done in the studio in terms of a boxed mechanism.

Before we open the floor to questions, I think in terms of everything that we have talked about tonight, there is an interesting conclusion that we can discuss briefly: that you have this amazing potential to interpret an artwork and the more successful an artwork is, the more people in a room can interpret it in different ways based on their experiences – their gender, their education, their loves and traumas. Numbers have the same potential, like with Gauss in the classroom – no one else could see the things that he saw and he found what you would regard as a shortcut. He looked at the numbers [and] some people may see a duration of time, in this number you would see a prime number; this could be the seven-day week, someone may say this is partitioned into a four and a three, some people may see a more numerological way as some sort of projected meaning of heaven.

We can open it to the floor; what does seven mean to different people? It is free to be as philosophical or as religious or as mathematical as you like.

I have got a final question for you, Marcus. Someone who I have spent a long time researching and making work inspired by her is Ada Lovelace.[17] She was this incredible visionary who is widely regarded as the first programmer, but mainly because she was the first person who saw that numbers could be more than just quantity, they could be entity. They could represent things that [were not] just apples and pears, they could represent data or they could represent information. She is the first person, decades, if not centuries, before other people, who had this amazing mind shift to see that potential of numbers. Then the last question I wanted to ask you all is this idea of a pattern that may have changed your life or changed your practice.

MDS Well, my area of research is trying to understand symmetry and what symmetrical objects are possible, not just in a three-dimensional universe, but in higher dimensions as well. I suppose if one thinks of just two-dimensional walls, what symmetries are possible: the Alhambra[18] is a great place to explore those patterns. In a way I am trying to understand what might be possible in high dimensional Alhambra palaces. Our first thought was that everything looked very chaotic and the way you put symmetries together seemed to be a complete mess. There did not seem to be any order to it and for me, the exciting moments [are] those moments of discovery. Seeing new things is what we live for. There are 'aha!' moments and for me it was actually understanding that there was a pattern, almost a Fibonacci-like rule, for how many of these things you will see. But I think the most exciting thing was constructing a symmetrical object that had a pattern hidden inside it that nobody expected before. The conjectures were telling us they were going to do one thing but as I built this thing suddenly there was a pattern emerging from a completely different area of mathematics, something called elliptic curves. Suddenly seeing this pattern inside this object has completely changed the direction of my research; it has opened up a completely new tunnel between these two. That pattern emerging, just seeing something appearing in a completely different area, for me, constructing that object was my greatest 'aha!' moment. If I was going to get a tattoo that would be [it]. Because I cannot draw them, it is not something you can see, it is something you can only describe in the language that we use to build symmetries. It would be a linguistic expression of this pattern that I am most proud of seeing.

17. Augusta Ada King (1815–1852), Countess of Lovelace, was an English mathematician and writer, noted for her work on Charles Babbage's proposed mechanical general-purpose computer.

18. The Alhambra is a palace and fortress complex located in Granada, Andalusia, Spain.

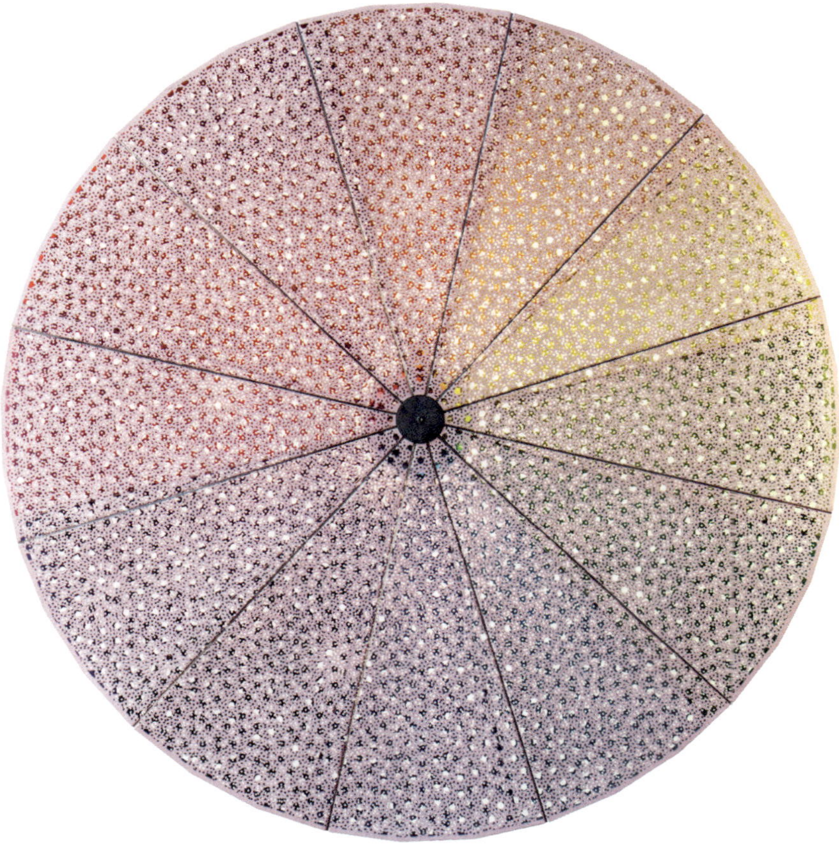

Patterns of Absence / Beacons

With the *Patterns of Absence* and *Beacons*, Shawcross created works that are akin to stained glass windows, activated by back light. Composed of two counter-rotating discs, each peppered with a myriad of complex perforations.

The *Patterns of Absence* are wall mounted, lit by a bespoke radiant light, whereas the Beacons stand on their own, activated by the sky behind them. The *Beacons* were installed in 8.5 metre form in Ramsgate, UK spelling the message 'HOME' in semaphoric marine code out to sea.

On the Relation between the Circle and the Triangle and the Ubiquitous Conical Form in my Work

Liliane Lijn

In 1959, when I took my first steps as an artist, I discovered the dynamic relationship between the triangle and the circle. I had made a drawing of a circle containing a triangle (*Two Worlds*, 1959). Inside the circle I drew a mass of creatures. The triangle remained empty.

I saw and still see the circle as a cosmic form, the potential container of life, a shape that represents the female body. The triangle, on the other hand, was a symbol for clarity, the clear luminosity of pure thought. However, the triangle was inside the circle. When I draw a circle, I imagine it expanding into three dimensions, turning into a cylinder or a cone. I can also open it and allow it to become a spiral whirling into a vortex. All these forms have become elements of my work as an artist.

In his *Greek Myths*, which deeply inspired me, Robert Graves[1] refers to the sacred white cone of ash that was called the White Goddess and was guarded by a priestess. The cone is endlessly interesting because, although it is built of ever diminishing circular planes, visualised in two dimensions it reverts to being a triangle. Cones are ubiquitous forms: steeples, skirts, space capsules, stupas, in our eyes they detect colour, not to forget mountains.

The following are a few examples of my work with the cone form.

Beginning in the mid 1960s, I began making kinetic conical works, some with poems or text Letrafilmed onto their surfaces in rings or ellipses, using fonts and spacing to indicate the tempo of their reading. On cork cones, I painted red or yellow stripes that then developed into the larger luminous *koans* that I am well known for, such as the *White Koan*, (1972) beloved of the students of the University of Warwick. An earlier example was *Anti-Gravity Koan* (1968) lost and remade as *Zero Gravity Koan* (1968/2005). It is a reinforced polyester white cone dissected into seven sections into which six planes of Perspex have been inserted. Four are tilted into elliptical planes while two remain circular. Internal fluorescent strip-lights illuminate the inserted Perspex planes that appear as rings of light embracing the conical body. Or the *Lost Koan* (2008) named after the two works lost in a fire in the collection of Mme Boissonas in the Cultural Centre in Flaines and the curious conical hole in our magnetic field over the North and South poles, where energetic solar particles are lost and thought to create the Aurora Borealis.

When the *Lost Koan* rotates, the luminous rings appear to move through the body of the cone.

1. Robert Graves (1895–1985) was an English poet, historic novelist, and critic.

It is especially interesting to observe these works outdoors as day passes into night, when the cone itself gradually dissolves, leaving only the dancing lines to indicate its presence. In 1974, I attended a seminar given by the theoretical physicist David Bohm and I was especially interested when he said that 'reality might be an interference system'. I was interested in interference patterns since the early '60s, and when, in 1982, I was commissioned to create a sculpture for a Science Park near Warrington, I called it *Split Spiral Spin: Project for a Sculpture as an Interference System*. My interest in interference that created moiré patterns led me in 1967 to design and make a *Kinetic Dress*. I liked perforated stainless steel, a material that was not at the time used much in design or art. The sculpture I designed is over 12 metres high, composed of eight parabolic curved sails, and is split apart in the middle. In plan, I drew it as a split spiral, the sails forming a spiral pattern. I intended the opening between the two halves of the sculpture to serve as a narrow path through which to walk and feel the energy of the interfering layers of steel. When seen with the sun behind the viewer, the work appears opaque, the stainless-steel sails reflecting light; however, when seen with the sun directly behind the sculpture, the sails are as transparent as glass. The interference of the four overlapping curves is very powerful, huge dark holes appearing to move across its surface as the viewer changes position. In 2019, I had a residency at the Laboratory of Particle Physics in Paris and then at Ego in Pisa under the direction of Stavros Katsanidis. For the final exhibition held in Pisa, I created a conical skirt that I set spinning, and called it *Gravity's Dance*. In some ways like the skirts of Whirling Dervishes in their lunar ritual, my conical skirt lifts into a wavy circular planar form. Once it attains the correct speed, the waves become more regular. Seen from above, the waves are three spiral arms. *Temenos* (2021) was commissioned by King's Cross Central Limited Partnership. It is over 11 metres high and eight metres diameter. Its open conical structure is made from 19 painted steel poles, the highest central pole supporting the nine poles on either side that successively decrease in height. A circular steel frame supports each pole at its base. The base circle is divided into 20 equal sections, the 20th section left open. The inside of each pole is fitted with LED strip lights that change colour from deep red to violet.

The design of this work is one of relation or connection between the separate poles.

Repetition has also interested me, because I found that repeating lines or forms could create energetic vibrations. I did a series of drawings I called *Flow Lines* (1972–76) in which I repeat the same mark, altering the angle of each subsequent mark.

I used this system for a three-dimensional sculpture commissioned by the Government Chemist in 1988. *Carbon Black* was made from a series of 48 identical folded triangular forms in painted perforated mild steel. When installed, the triangular elements created a helical form that arched from where it was fixed to the floor to a wall approximately six metres away and four metres high.

I used identical geometric elements to create an organic form dependent on the spatial relationship and connections between the separate elements. I was told by the Government Chemist that it seemed to be a very similar process to the organisation of some naturally occurring molecular structures such as the double helix of DNA.

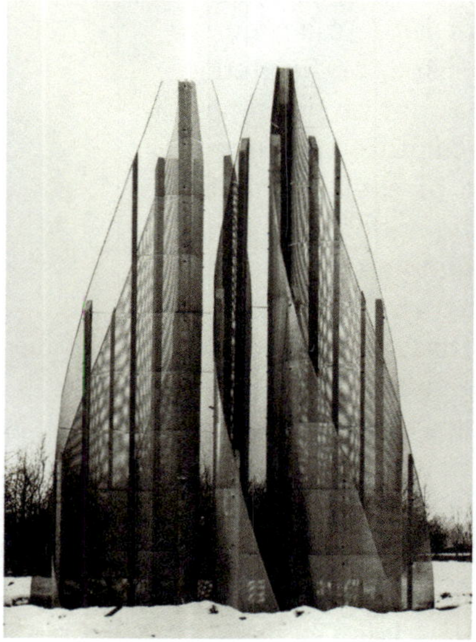

Liliane Lijn, *Split Spiral Spin*, 1982

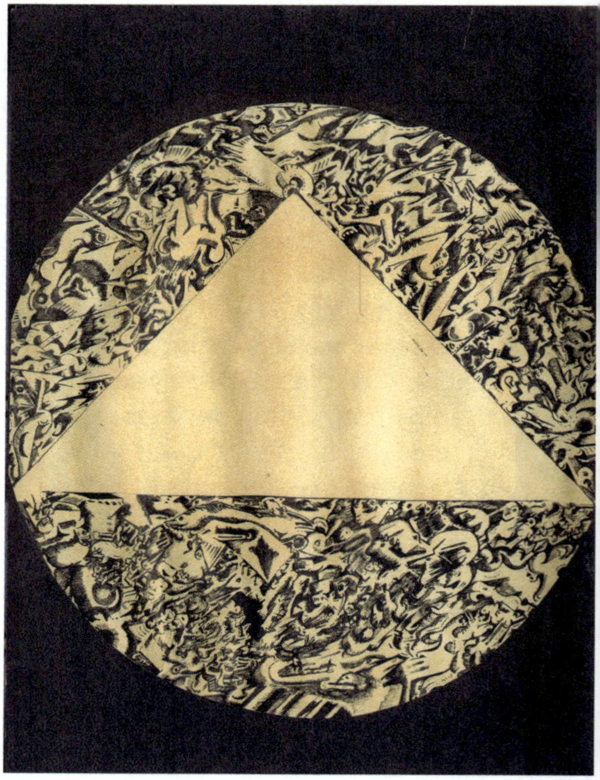

Liliane Lijn, *Two Worlds*, 1959

Memory: the Mathematical Mind and the Art of Pattern Searching

Marcus du Sautoy

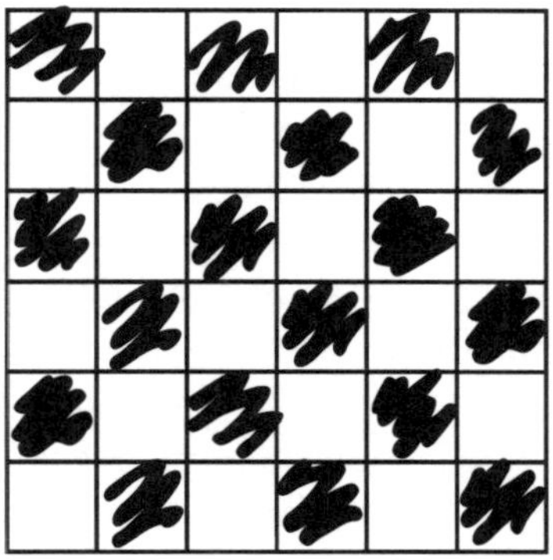

1

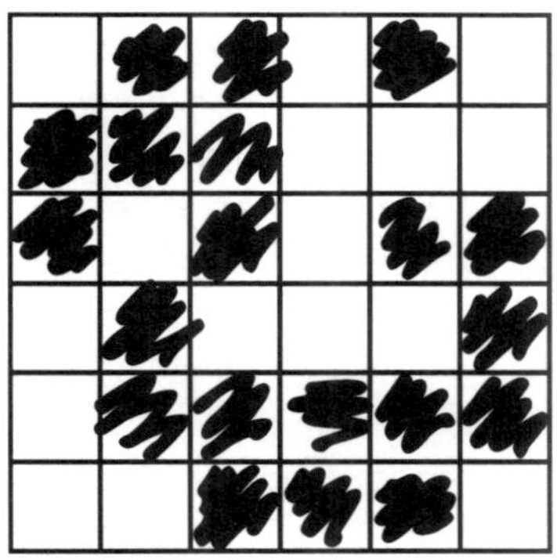

2

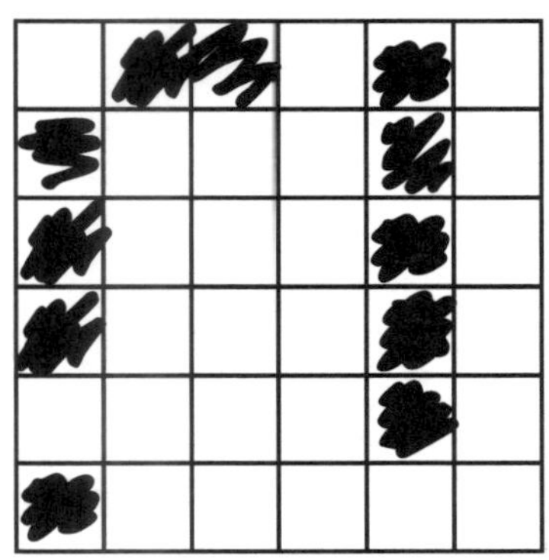

3

Stare at the squiggles contained in the 6×6 grids. Then close the book. Which of the images can you reproduce from memory? The key is not to try to remember each of the 36 squares in the image individually but to find a pattern that helps you to generate the image.

This problem goes to the heart of what it means for me to be a mathematician. A mathematician is a pattern searcher, someone who tries to find logic and structure in the chaotic world that we inhabit. For me it is key to the way that we memorise things. Memory depends on identifying pattern and structure to help our brains store a condensed programme from which to regenerate the stored memory.

Mathematics has always been my weapon against my terrible memory for names, dates, and random information that I can make no logical sense of. In history I have not a clue what date Queen Elizabeth I died and if you told me it was 1603 I will have forgotten it ten minutes later; in French I always had difficulty recalling all the different forms of the irregular verb ‘aller’; in Chemistry was it potassium or sodium that burns purple? But in mathematics I could reconstruct everything from the patterns and logic I had identified in the subject. Spotting patterns replaced the need for a good memory.

Take, for example, the 36 squares coloured alternately like a chess board (Image 1). It is easy to remember. The pattern is obvious. A simple algorithm suffices to generate the whole image. Faced with a grid that was arbitrarily large the same algorithm would help the brain to encode the memory of the image.

Although the second image has roughly the same number of squiggles the lack of an obvious pattern makes it much harder to remember. The image was generated by tossing a coin and marking a square with a squiggle if the coin landed heads. Mathematically the coin is equally likely to produce the first pattern of squiggles, alternating between heads and tails, as to produce the second arrangement of squiggles. Yet we would identify the second image as more random.

This idea of the size of the algorithm needed to memorise an image is a powerful measure of the randomness contained inside the image. The chess board pattern is very ordered. Its generating algorithm is small. The image generated by the toss of a coin probably requires an algorithm that is no shorter than recording the contents of each of the 36 individual squares in the grid.

Identifying whether a process is random or has an underlying logic controlling its development has been crucial to our survival from the time humans first evolved in the African savannah through to our trek through the modern urban jungle. This facility that the human brain has evolved to search for patterns is a deeply mathematical one. Despite most people’s knee-jerk reaction to claim they are bad at mathematics, this hunger for pattern and the efficient memory making machine that our brains have become for me is evidence that we all have mathematical brains.

The third image is a good measure of how good you are at pattern searching. There are not as many cells to remember as in the first two images but enough to be just outside our memory comfort zone. At first sight the arrangement of squiggles has no clear geometric pattern that provides access to memorizing them efficiently. Yet looked at in the right way it hides a simple mathematical meaning that will generate the images.

Write down the numbers of the

squares that are coloured and we get 2, 3, 5, 7, 11, 13, 17, 19, 23, 29, 31. These are the prime numbers, the indivisible numbers. On one level this certainly provides a rule for reproducing the entries of the 6 × 6 grid. But what if we start to enlarge the grid? Suppose I extend the image to 1,000,000 rows and six columns. What algorithm would you apply to reproduce the prime grid?

One observes that beyond the first row the second, third, fourth and sixth columns will never contain squiggles because their grid numbers are divisible by two or three and so will never be prime. But what about the challenge of reproducing the squiggles in the first and fifth row?

It turns out that these are as difficult as remembering a random sequence of squiggles chosen by the throw of a dice. The primes are the ultimate challenge for the pattern searcher because outwardly they look so random with no discernable pattern. In fact this search for an algorithm that will help us 'memorise' the primes is the greatest enigma on the mathematical books. Although we know that the sequence of squiggles continues infinitely, the largest grid entry that we know contains a squiggle is the square numbered $2^{82,589,933}$-1.

It is this tension between order and randomness, between memory and forgetfulness that makes the primes the most interesting part of the mathematical gallery. The same tension is at the heart of those visual patterns that excite and surprise. From the strange tiles that adorn the Alhambra to the abstract designs of the contemporary world, the way our brain responds to these patterns is based in part on our innate mathematical desire to seek out an algorithm or rhythm that helps us to navigate and memorise the design.

From humans to machines, anyone or anything that is memorising something is applying a distinctly mathematical side of their brain. Memory requires finding patterns, connections, associations, and logic in the data that we are trying to store. Whether we are aware of it or not, we are all mathematical pattern searchers at heart.

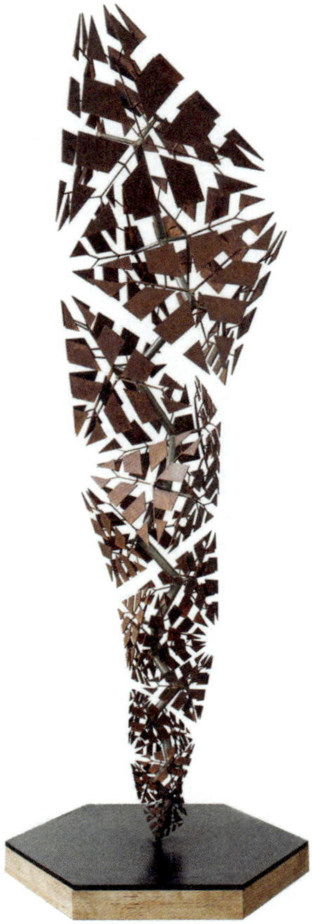

Fractures

Fractures are ethereal forms that almost seem to disappear as they rise up and expand. Imbued with references to the manmade and natural worlds, the *Fractures* contain a central helical stem, a twisting spine that supports a series of branches which in turn support hundreds of triangular leaf-like fragments that dissipate in precise formation. The pieces grew from the initial explorations of the *Paradigms*, exploding the form into triangular shards. A 4.5m memorial to William Butler Yeats entitled *Enwrought Light*, with silver and gold leaves, was installed in 2022 in Chiswick, London.

Lecture Four, 9th November 2023: Forming a Thought into Form

Speakers: Jon Keating, Maya B. Kronic, Emma Ridgway, Conrad Shawcross
Moderator: Fatoş Üstek

Fatoş Üstek

Could you perhaps remind us what was your thinking, the initial inspiration, when you saw the works in the exhibition and what it evoked in you?

Jon Keating

Thank you. Well, thoughts form and they also evaporate, so I do not know exactly what I had in mind but I think I can reconstruct it. I was seeing the sculptures being placed and it encouraged me to think about the role of the observer, the beholder, in art as compared to science. I am sure others are better qualified to talk about the role of the beholder in art and that people who produce art think about the impact that will have on people and that therefore influences what they produce. I can imagine that is especially the case in performance art, where sometimes the beholder is part of the artistic production. But in my mind, I was contrasting that with the role of the observer or the beholder in science.

Up until about 1920 people thought that objects in the world, like a sculpture, were there whether or not there was anybody there to see the sculpture or not. Objects in the world have properties and these properties could be measured. Some properties might not have been measured and then we are ignorant of them, but in principle we can go and measure them, and they exist in that state before we measure them. In about the 1920s, it was realised that at the smallest scale in the world that is no longer true. Quantum mechanics is an inherently probabilistic theory of matter and it is a very strange theory. It has an inherent randomness to it and you cannot really speak about something having properties before they are measured; the act of measurement produces the property and that is an inherent part of quantum mechanics.

Some of our audience may have encountered the Schrödinger's cat experiment (not a real experiment I hasten to add) where you put a cat in a box and a quantum process which is inherently random can fire off some poison which would kill the cat if it is fired off. If it is not fired off the cat will live and so if you are outside the box and cannot look in, you do not know whether the cat is alive or not. The cat is in some quantum superposition of states; it is both alive and dead and it is only when you open the box that the cat either is alive or it is dead. It is the act of observing that forces it into that state and this was deeply troubling to people early on in quantum mechanics, in particular Einstein who famously questioned whether this could possibly be true. He had an intuition that objects did have an intrinsic reality whether or not they are observed and he famously asked: does the moon exist if no one has actually observed it? In quantum mechanics the answer is no, but that is not our everyday experience, and so I was merely looking at the sculptures and thinking: there is a quantum view of this, which is: it is either there or not there. If I have not looked at it, before I have looked at it I do not know whether it is there or not: it is in a superposition of states.

FÜ That is quite interesting because also that sculpture is very heavy, so the tangibility of the sculpture mostly feels like it is there. Jon, thanks for the brilliant revisit of that crucial moment (for me at least). Could you perhaps expand on your reception of the title *Forming a Thought into Form*, in the context of quantum chaos and the random metrics theory which are your specialisations? It would be really pleasurable to hear about your interest in number theory in this regard.

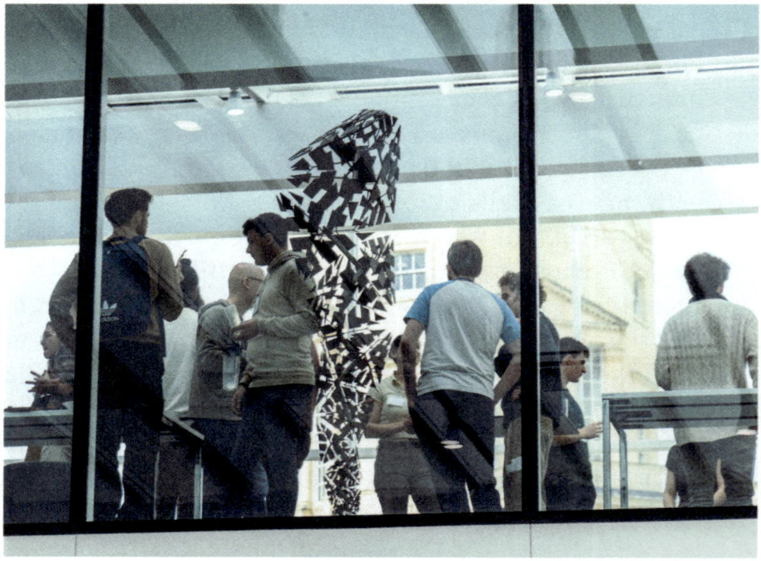

Fracture (R24H12), 2019

JK I wanted to give an even more general response than that, which is the response of a mathematician: how do mathematicians proceed from a thought to a form and what is a form? For us it can be many things. I will say a little more about this later, but it might be an equation, it might be a quantitative understanding of some process in nature, it might be a question. I wanted to say a little bit about what mathematicians do, what they like to do, and what they are aiming towards. Mathematicians do like to think, thoughts are very important to us (as they are to everybody). Mathematicians are particularly wedded to thinking and can think about a particular problem for days, weeks, months, years, decades. There are some problems that I have thought about ever since I was a student and I think about them most days, certainly most weeks, and I can think about them for a few hours if I am driving along. Thinking is an essential part of the mathematical process, and it is a very creative part because one is really trying to explore the possible in one's mind. We very much like to verbalise, surprisingly. Mathematicians are thought of as being introverts but when we are with other mathematicians we are not, and a key part of converting a thought to a form for us is talking about it. We meet for coffee every day, we discuss, and that process of sharing is critical to developing the form that we produce. We also write, we sketch, and we doodle. It is very rare that you meet a mathematician who does not have a piece of paper and a pen, and that is how we communicate. We talk and we write at the same time and that combination of words and writing is critical. Actually, these tables are designed so that you can write on them with felt pens and that is what we do over coffee: we talk and we share ideas by writing, we visualise. We like to use our computers to draw pictures, sometimes models of images that we really cannot even comprehend, so these might be shapes in a very high dimensional space: in 1,000 dimensions, 1,000,000 dimensions, an infinite number of dimensions. One cannot really produce sculptures or models of that, but one can try to draw pictures that capture the essence of that. That is very important to us.

When we develop a form, we do not just stop there; we can polish it for again days, weeks, months, or years before we are prepared to release it, because we want it to be perfect. Once we have had an idea and developed a form, we are very comfortable thinking: 'I would like to have another go at that now, do it slightly differently.' We generalise, we work from special cases that we understand, and then we use that to form general theories that we hope are true, that we then try to explore. I would say we collectivise in that we do not feel we have ownership of thoughts and forms; we share them with other people. We are very comfortable that we might have a thought and that somebody else may then come along and take that thought forward and produce a different form. We experiment and we challenge, we criticise each other, we criticise ourselves, and that is a key part of the process.

FÜ This is an amazingly clear outline.

JK This is now what we [are] aiming for. So, what do we do? We formulate questions which are very important to us, and a key element of research might be to simply ask a good question. We make guesses, we of course solve problems when we are able to. The key point is making connections, seeing connections between different areas of mathematics, seeing that they are related, and that and unexpected connection is tremendously important to how we form new perspectives.

We are comfortable using a word, which I think many people in the in the art world are not comfortable with. We try to produce things that are 'beautiful', and we are unselfconscious about using that term. We try to produce beautiful proofs, we aim to understand, we aim to discover, we want to be increasingly precise, so mathematics we think of as being ultimately correct or not, and we want to get the most correct description of the world. We are conscious that there would be something better but we want to produce that, and I guess finally we want to be right, and this is where there is maybe a difference between the art that Conrad and others produce, and what we do, in that Conrad has the luxury that he can produce whatever is in his imagination. We are constrained by being right about something. We can be wrong, we can produce something and someone finds a mistake in it or says that does not describe the world if we use your equation, it is inaccurate. So, we have the work behind us, this constant fear or hope. The hope is that we will produce something that is correct and the theory is that we will not.

I will finish with two analogies that are commonly used. Doing mathematics has been likened to walking into a room that is completely pitch black and you do not know what is in the room or where anything is, and you are groping around and do not know where to go. There is nothing that leads you in any one direction, but slowly you form a feeling around a sense of what is in there, and then you find a light switch, and suddenly understand what is in that room. Then you move onto the next room. Or climbing a mountain for the first time when you do not know the right route and you are constantly searching for the best route, the optimal route, the safest route, the route that gets you to the top, maybe the most beautiful, aesthetically pleasing route. That is what we aim to do now.

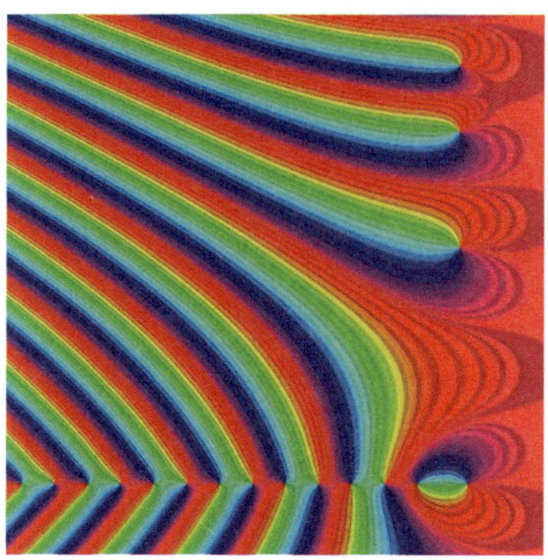

A phase portrait of the Riemann zeta function

This is an example that I wanted to give of a form; this form represents something that is very important in mathematics, called the Riemann zeta function. I do not need to tell you what that is, just look at that picture and you see the colours. Think of those as contour lines, think of this as a landscape with contour lines. We are very interested in the points where all the colours meet. Those are tremendously important; they are called the zeros of the Riemann zeta function and if you walk vertically up through that line of zeros you see a picture like this, which you might imagine as walking on a landscape along a ridge of some hills. You go up and you go down and we would like to understand how high you go and how low you go. It has been a problem for 160 years in mathematics to understand the shape of this ridge of hills. There has been tremendous progress in the last ten years by taking a formula that we have known for 160 years and realising that it has a structure that we never expected. Now when we see this formula, we cannot look at it and see anything other than this structure, but before we understood this ten years ago, we had no idea that that structure exists. That structure is actually perfectly represented in one of Conrad's sculptures which I am going to show you later.

FÜ Thank you Jon. Maya, can you talk us through your approach to producing a thought, that is again that relationship between form and process?

Maya B. Kronic

In the presence of my illustrious co-panellists I feel like the first thing I have to say is that this is all a huge mistake. I have really [fallen] into the profession of being a publisher and an editor totally by mistake and now, 15 years later, I am still the Director of Urbanomic,[1] a struggling, small publishing company. My background is in philosophy but what I have tried to do with Urbanomic from the beginning is to publish work which is on the edges of philosophy and abuts, or crosses over, or communicates, with other spheres (philosophy and art, philosophy and science, philosophy and culture, and music). All this happened because in 2007 I decided to publish this journal *Collapse*. One way of describing this, if I were to try to psychologise myself, would be to say: here is a person who found they were profoundly allergic to specialisation and professionalisation of knowledge, something of a dilettante and a serial enthusiast with a personal dissatisfaction with the academic framework of philosophy.

After entering university multiple times, I finally decided to turn my dissatisfaction into a kind of personal project. It was really accidental, but it was driven by two motives, one negative and one positive. The negative one is that I find that university philosophy, academic philosophy, was a very constrained discipline which tended not to communicate very much with other practices and modes of thought. The positive motive is that I think everyone encounters philosophical problems, usually when the established conventions and regulations and rules of whatever your sphere is, do not work anymore. I think everybody eventually, now and again, encounters a philosophical problem, and I wanted to provide a place to do

1. Urbanomic is a British publication house founded by Maya B. Kronic.

philosophy in the way that I understood it, as a kind of inventive, conceptual thought that happens in all sorts of places when things break down.

I wanted to bring together people writing in different registers, authors from different worlds, different disciplines, and published writers, also philosophers who would not necessarily find a voice in the 'academy' as it was. *Collapse* was conceived as bringing the eclecticism of a magazine together with the serious intent of an academic journal.

Usually, I would think about a journal as being a space which has a clear demarcation [or] circumscription of the kind of content, the criteria, for what could be included in it. At a higher level this communicates with the university model of knowledge as a whole. University obviously means you have this universal model of knowledge within which you have disciplines and then subdisciplines. There is a kind of structure of knowledge that is presupposed in that very idea, in the way in which universities carry out their business. Of course, now we have a lot of interdisciplinary work, but what does interdisciplinarity mean? It is a very contested term; it often serves just to introduce an interface between one subdiscipline and another, without really disrupting the circumscription of the disciplines themselves.

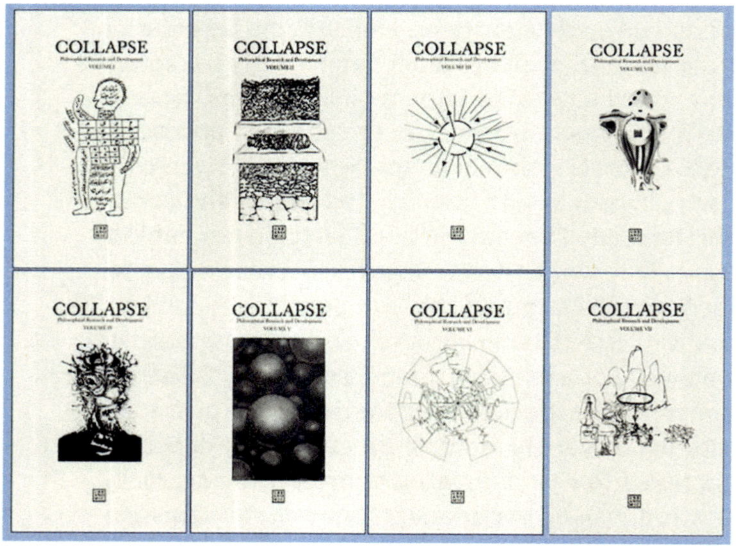

Covers of volumes 1–8 of *Collapse*

My idea for *Collapse* was to be undisciplined and transversal, so unlike academic journals where all the contributions are circumscribed within a subject area (that circle is then contained within these concentric circles). What I tried to do was to create a kind of montage with artists, scientists, theorists from different disciplines, people writing fiction, artists, with each contribution overlapping slightly with the next one. You end up with this weird connected chain where the two extremities of the chain might have two contributions that have absolutely nothing to do with each other, but they would be connected by these intermediate links. That really became my editorial work, putting together other people's thoughts.

For instance, my favourite volume (I think volume 7, on cookery) talks about materialism, cookery. It has theorists, philosophers, anthropologists, it has articles on the history of industrial agriculture, the use of chilli as a weapon, the meaning of synthesis in mathematics, the artist Gordon Matta-Clark's[2] art café, anorexia, how the discovery of cooking food helped define the evolution of the human brain, predation and cannibalism, and the link between pigs and whey powder. The way I try to think of it is as a different topology of the concept, and one that I think is more contemporary than a hierarchical delimitation of specialisms. It allows us within one volume to trace out the multi-dimensional space of a concept, to take that concept through all these different worlds, and to produce this kind of montage, a journey. Now, the second interesting thing about that is it also assembles a new audience, because people will often pick up one volume knowing one of the authors and they get introduced to all this other weird stuff. It has actually produced an audience for our work that probably would not have existed.

To come to the question of form, if what we are doing is trying to connect together all these things, why not just do it on the internet? You have hyperlinks, you can link everything together, you can create this sprawling web of knowledge. A few reasons: one is my self-worth as an editor. I need to think I am doing something and I think the editor is like a kind of lens, a concentrator. I have my particular interests, my history, and the people who I have encountered. You can go away and produce your own journal and yours will be different from mine. I think there is a kind of focusing function. The second one is sequencing: in a book you have a linear sequence and as an editor you are guiding people through this weird journey through different disciplines. The most important one for me is commitment; it is that when you put something into a book form, you are making a statement. You are saying: 'I have put these things together; you can think I am crazy or stupid or wrong, but I have done it, it is a commitment and I am going to publish it'.

Things have moved very fast since 2007, so where are we with the book form today? When I printed *Collapse* one, I think I had to print 1,000 copies, because that was the only kind of viable number. Nowadays of course you can just put an eBook out. We are selling more and more eBooks, but also you can do print on demand, of one copy of the book. The progress of technology is changing the nature of this venerable book form and

2. Gordon Matta-Clark (1943–1978) was an American artist noted as a pioneer in the field of socially engaged food art.

what it means to publish, to form thoughts into this format. These older forms like books may retain qualities and advantages that newer forms do not have, but forms themselves do not remain stable. When put into a form, thought is still not fixed.

Appropriately for this occasion, I met Conrad through doing *Collapse* and he is featured in several volumes. We actually ended up doing a project with Conrad where I put him in a room with a mathematician for a week and we talked about the zeta function.[3] I think of all this as being really a sculptural thing. When I am putting this together, I feel like I am squeezing things together, finding how the shapes can possibly fit. In the end I want it to have its own integrity and I want to commit to it, but it should not be self-contained. The form that I like for editorial work is that it has an integrity, but it also has strange spiky bits pointing out in different directions. It is not smoothed out; it comes with all these bits from other worlds still attached to it. That is my inner vision of the form that I am trying to create as an editor.

FÜ That is very, very, beautiful. Thank you so much, Maya. We were talking on the way here with Conrad: it is also about proliferating further thoughts through encounters, and those encounters also project onto new meanings, because everybody comes from a different background, different set of perspectives of seeing, engaging, relating to the world. Introducing this unexpectedness into a contextual engagement is quite inspiring, as you said. It also not only proliferates the profile of audiences, but also it creates that playfulness, or the potential of coming up with something that you would not come up with. It is almost like unthought can be attended [to] for a split second of time before it flies off.

Conrad Shawcross

I loved your use of the lens as an editor. It actually reminded me of Brian Catling,[4] one of the ways he described his process of teaching. Brian was a tutor of mine at the Ruskin and he was very much part of the reason this whole show came about, this idea of taking people outside their comfort zones, putting an artist in a room with a mathematician, which sounds like [I was] a wild dog but it was not quite that bad. It is really about this kind of export and import, taking artists into the Maths Institute, and mathematicians into an art context, really outside one's comfort zones and seeing what happens. Brian described his role as a tutor as holding a distorted mirror up to his students, so it was again this lens idea, not preaching or being some kind of guru, but allowing that person to see themselves.

3. The Riemann zeta function arises in definite integration and is intimately related with very deep results surrounding the prime number theorem.

4. Brian Catling (1948–2022) was a British sculptor, poet, novelist, film maker, and performance artist.

FÜ Thank you, Conrad. Emma, thank you so much for joining us today and many congratulations on your new role as the Director [of the Foundling Museum]. As I informed our audiences, we have been in touch [for] almost two years now, and you have been an avid supporter. We were exploring ways of collaborating with you while you were at Modern Art Oxford and we were very much inspired by the exhibition that you curated of works by Ruth Asawa,[5] who is a deceased artist. Her work came to prominence over the last years. Can you perhaps introduce us to Ruth's body of work, but also your curatorial approach in shaping the exhibition? Perhaps my question is a bit more practical, more tangible, now that we have come through the steps of mathematical, editorial, philosophical thinking.

Emma Ridgway

Thank you and thanks for the invitation to join you here. The [first solo show in Europe] of the works of Ruth Asawa was titled 'Citizen of the Universe'. I took that title from a line I saw in a love letter she wrote in 1948, as I was exceedingly taken with that idea. I co-curated the exhibition with Vibece Salther of the Stavanger Art Museum. Asawa's work and her life were really unusual and one of the reasons you will see when you see the work. Ruth Asawa was born in 1926 in California; her parents were Japanese immigrants. (Her mother actually never learned to speak English.) So, Ruth was brought up in a very interesting moment of history. I sensed that she had a special kind of joy that she arrived at mostly through her experiences at the extraordinary education institution that she went to, which was called Black Mountain College.[6]

If we look at the photograph of the Asawa exhibition in Oxford, we can see a number of her sculptural works installed together in one large gallery. They are exquisite – really extraordinary. In terms of my curating, I trained as an artist; I approach life and my work in a visual and embodied way, through physiological experiences. One of the things I love about exhibitions is that you move through physical spaces. So, I approach curating exhibitions as choreographing visitors in terms of how they are likely to move through the spaces. I usually spend a long time distilling down the key ideas.

With the texts in exhibitions, I distil down lots of research in order to highlight just a few key concepts, to invite visitors to focus their attention initially on a few ideas; such that people can take a hold of those and explore the space and the ideas, on their own terms. I am really fascinated with Asawa's work because it made me gasp when I first saw some of it in

5. Ruth Asawa (1926–2013) was an American modernist known for her abstract looped-wire sculptures.

6. The artistically influential Black Mountain College operated from 1933 to 1957, in North Carolina, USA, and taught a number of artists who went on to become very successful.

MoMA New York. A wire sculpture initially looks like it is just born of natural materials and then you start to realise it is handmade. The more you see, the more extraordinary it is, because each work is made from single lines of wire. There is no welding, there are no fixings, they are looped into themselves. She has two distinct techniques: one is looping wire and the other is tying wire; each drawing inspiration from how nature repeats rules to generate and grow forms.

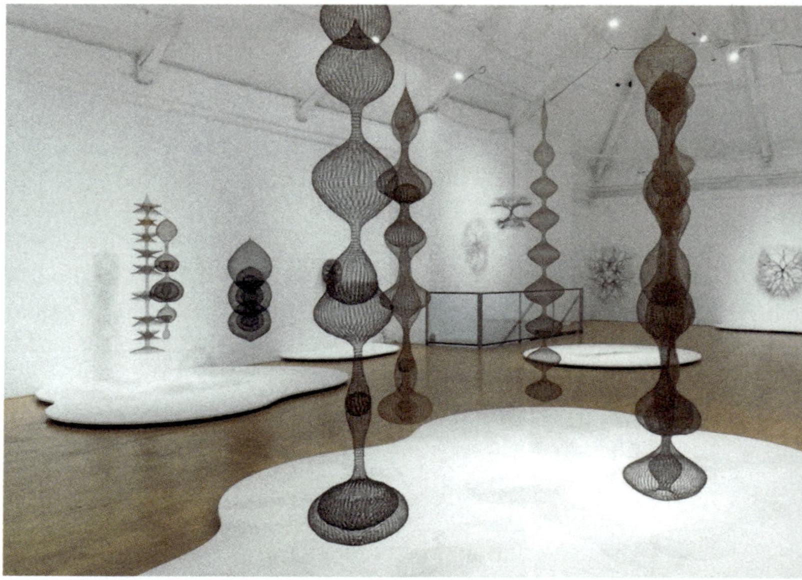

Ruth Asawa, *Citizen of the Universe*, installation view at Modern Art Oxford, 2022

Some of her hanging looped wire works are huge, over three metres tall. They are suspended by their own weight. Importantly, they are some of the largest volumes you can manage on single lines of wire in which the material itself is just holding its own [sculptural] form intact. She came to this very distinctive individual language of sculpture when she was at Black Mountain College. Black Mountain College [BMC] was a progressive college that was exceedingly unconventional. Joseph Albers[7] and Annie Albers[8] were invited over from the Bauhaus[9] in order to lead the programme. BMC was an interdisciplinary college very much based on ideas of John Dewey, [a] progressive educationalist who believed in whole person education and coming at education in an interdisciplinary way. The Albers put art education at the centre of the studies at BMC. Every student had to [approach] art as a core way of exploring and engaging with the world, with the training by Joseph Albers as the foundation.

7. Joseph Albers (1888–1976) was a German-born American artist and influential teacher, who was known for his approach to colour theory.
8. Annie Albers (1899–1994) was a German textile artist and print maker, credited with blurring the lines between traditional craft and art.
9. Bauhaus (1919–1933) was a German art school that combined crafts and fine art.

He was brilliant at training people *how to see the world* as an artist.

There are images of his wire exercises [that] he would get students at the Bauhaus to do, which went on to have a huge influence on design. As with the exercises he taught at BMC, the purpose was to explore how you could use simple materials to make a volume of space. What he was intending to teach was something we still get taught in art school: how can you sharpen your visual perception by shaping lines around 'negative space' (either drawing or with simple materials)? How do you start to switch your visual perception so that you move into what some call a right-brained mode? It is exceedingly wordless. It is such that you do not see 'a bottle' there [speaker points to bottle on the table], because that [idea of an object] would be really hard to draw. Instead, you get into a zone whereby what you are perceiving contrasts in negative and positive space. By being in such a different headspace you are able to create spatial and visual things such as drawings or sculptures. Ruth Asawa was also very influenced by R. Buckminster Fuller [although] she was not actually in his class. R. Buckminster Fuller, some of you will know, [was] very experimental, super charismatic, super eccentric, and interdisciplinary; he wove together lots and lots of different ideas with great energy.

[In] this period, from the '20s on, there was far more influence [from] Eastern thinking for various reasons, [Fuller was] part of this move against [conventional Western] ideas of static, certain knowledge, and more influenced by ideas that the world is a continually changing and evolving set of living things. In that process-based way of thinking failure was a way in which you could try and find new forms. The geodesic domes we are all familiar with started as this failed experiment in a summer school at Black Mountain College. Fuller brought the ambitious idea (but no plan) for a lightweight dome and Venetian blinds to build it with. The premise was to use the least amount of material to make the most volume of space. (So, you can get a sense where Ruth Asawa drew that idea from in her work.) Fuller and his team completely failed on their first attempt, but they improved the design and succeeded the following summer. For Asawa, other teaching that was key in this extraordinary college [was] again different ways of making physical forms.

On the right we see what becomes Ruth Asawa's drawn motif, which later informs the shape of her sculptures. But at this point, as a highly talented young artist, she is expected to be a painter. I love the photomontage of Asawa dancing or flying, made by her friend in John Cage[10] and Merce Cunningham's[11] dance class. They were very fascinated by Buddhism.

10. John Cage (1912–1992) was an American composer and music theorist, he was a pioneer of indeterminacy in music, electroacoustic music, and non-standard use of musical instruments.

11. Merce Cunningham (1919–2009) was an American dancer and choreographer who was at the forefront of American modern dance.

So ideas of dance and movement from Merce Cunningham, for example, were not about synchronicity in space (as we might expect in a traditional ballet, for example). Instead, this dance class was about how you make forms with your individual body, such that it moves through time. Time [was] marked by Cage's abstract music scores. These unconventional ideas deliberately set out to open up one's understanding of what it is to be an individual in the world in relation to others, and in relation to something larger than yourself – the group, time, being alive right in this moment. To [go] back to your earlier question, that is what drives my curating. What I am fascinated [by] is exploring different perspectives, different ways of understanding being alive through different individuals' subjective perception of the world, and how we can learn from that and apply that to our own understanding.

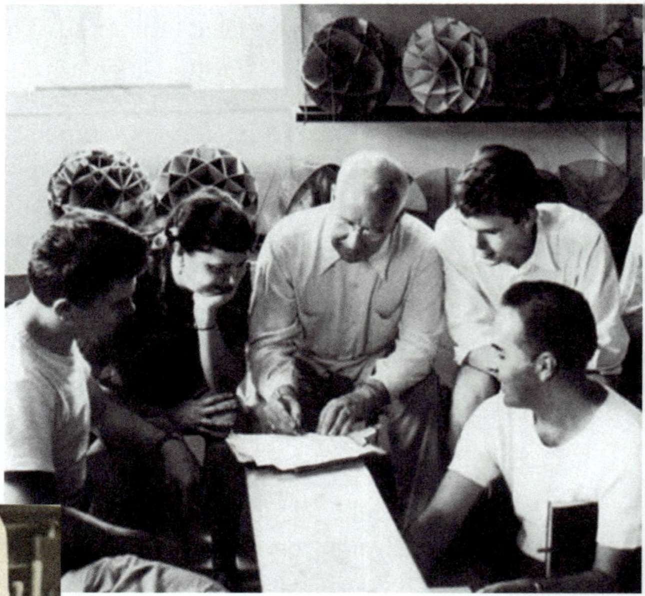

Buckminster Fuller and students, Summer Institute, Black Mountain College, 1949

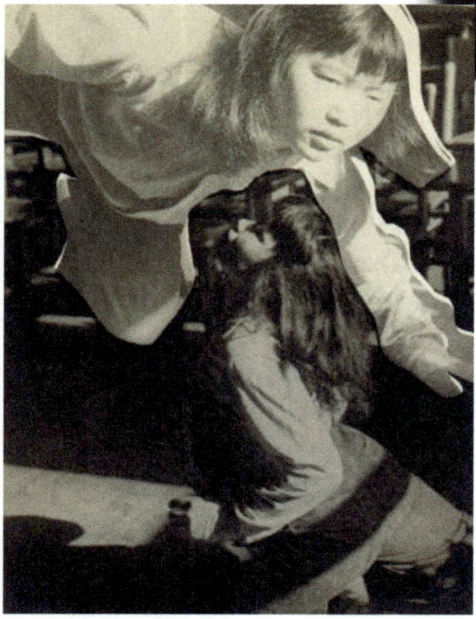

Collage of Ruth Asawa Portraits at Black Mountain College, North Carolina, c. 1948–1949

The best-known photograph of Asawa shows her sitting on the floor, looping wire to make her artworks, surrounded by her children. The exhibition title 'Citizen of the Universe' I took from a love letter she wrote nine years earlier to a student at BMC. They married and they went on to have many children and an impressive life together. She continued to make her work all her life and also did community projects separate from her own individual art practice. Her ingenious way of creating works with simple wires meant that she could work at home whilst also being a mother, and she [was] exceedingly community-minded. Asawa was absolutely extraordinary.

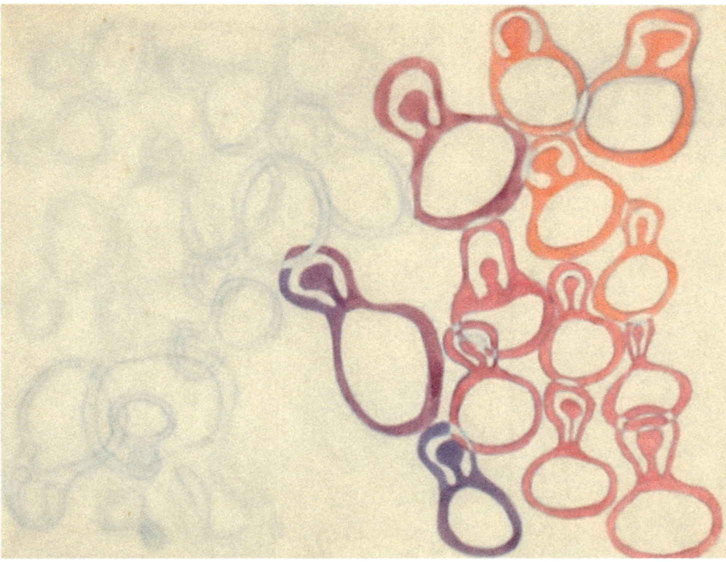

Untitled (BMC.144, Dancers) c. 1948–1949

Asawa's community work with schools shows her approach to teaching, which was influenced by Fuller's methods at BMC. Her and an art historian friend, and later some neighbours, set up an elementary school arts education programme because they felt it was so poorly done by the schools. Asawa and her collaborators brought in professional artists, professional musicians, and others to teach.

The Big Circle is made with paper plates and hair clips and it is a way of folding and piecing it together. The story that goes along with it – as it is being put together – is a philosophical idea of us each being parts that complete a whole, part of the big circle of life and the universe. Ruth's understanding of mathematics [was] about being something that is within the world to make sense of one's place in the world.

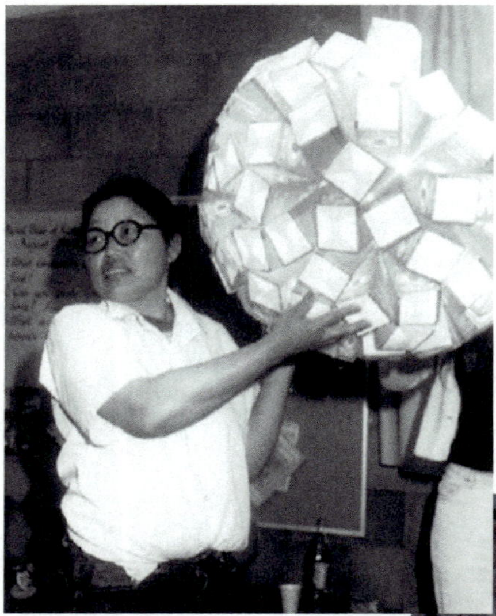

Ruth Asawa holding up a milk carton construction of a globe at McKinley Middle School in Redwood City, California, 1972

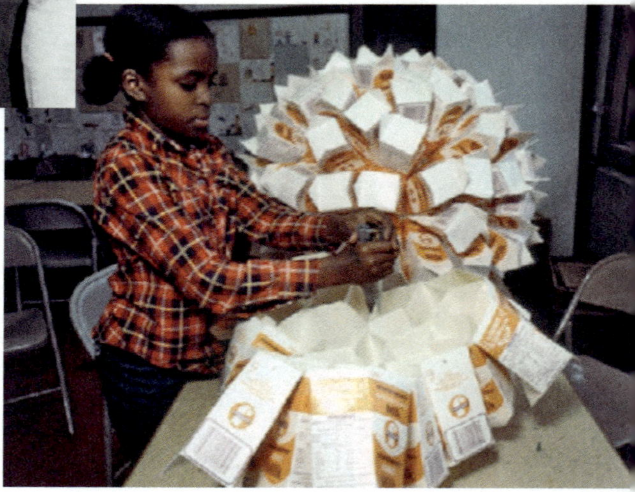

A student using milk cartons to make a sculpture, 1981

Ruth comes up with this way of using recycled milk cartons (that are quite hard to get in the UK). It was a children's rights and children first moment in the '70s and this is California – all the projects were very orientated around how you encourage children to work together to create something larger than the whole, something they can work on collaboratively. She and Bucky came up with these brilliant milk carton constructions and her daughter Aiko worked with her (and others too).

So, there are a few things that are really striking; like the 'Citizen of the Universe' concept that I was so taken with. I of course did that [research] rabbit-warren thing you do when you are trying to make sense of where she [got] ideas from. I think the place that it probably comes from is Leibniz.[12] At that time Bertrand Russell[13] had written a very popular book for students to understand philosophy and there is a

12. Gottfried Wilhelm Leibniz (1646–1716) was a German polymath, perhaps best known for developing theories around integral and differential calculus.

13. Bertrand Russell (1872–1970) was a British mathematician, philosopher, and logician.

lovely, slightly simplified articulation of Leibniz's idea (which I do wonder if she came across):

> Contemplation enlarges not only the objects of our thoughts, but also the objects of our actions and our affections: it makes us citizens of the universe, not only of one walled city at war with all the rest. In this citizenship of the universe consists man's true freedom, and his liberation from the thraldom of narrow hopes and fears.

She talked about her tutors (including Max Dehn)[14] as having taught her and profoundly influenced her outlook on the value of life and the true nature of freedom. The idea of a citizen of the universe offers an identity beyond your immediate mother-daughter-immigrant roles and narratives.

[The Milk Cartons] class they then made into something that could be rolled out across elementary schools. It was [an] incredibly successful workshop programme. In the 1981 introductory pamphlet of how to teach this for maths or art in elementary school, I was so taken with this succinct articulation of forming a thought into form. It opens with this description for teachers:

> The usual approach for geometry in school is to start with plain geometry and then proceed to solid geometry; however, the three-dimensional world of solid objects is much closer to the student's experience.

Bear in mind John Dewey[15] was all about learning from experience and experience is how we engage with the world. The British mathematician, Grace Young, had already pointed [this] out in the 19th century so I of course looked into her. She was writing brilliant books on how young people can understand the world if you make, with them, paper models of geometry and other things. They can really start to engage with abstract ideas which resonates with their experience too.

To go back to Ruth, by dealing with the objects the child can touch and move and even toss around; more of the senses are involved and learning takes place naturally and in the spirit of fun and exploration. This approach to learning has been discovered and rediscovered many times. People have worked with art and mathematics to mutual benefit in both fields.

I will finish it there but there are lots of interesting resonances between the multi-disciplinary and inter-generational learning and passing on which I know resonates with Conrad for Brian Catling.

14. Max Dehn (1878–1952) was a German mathematician, known for his work in geometry, topology, and geometric group theory.
15. John Dewey (1859–1952) was an American philosopher, psychologist and educational reformer.

CS It reminds me also of another pioneer in her field: Maria Montessori[16], and her coining of the phrase 'psychogeometries', and that wonderful idea of how geometry has all this potential. That was a thing she did in preschool, in nurseries, before there was language to promote neural pathways and ways of thinking. She would [make] sequences of colour, of geometry, of form, repetitions like this; that arrival at complexity quickly through the repetition of something simple is I think at the core of a lot of what we are all talking about here (and within nature). It is also then trying to extract those simple cores from the complexity we have around us: getting back to that kernel, that essence, that atom.

FÜ Thank you so much for all your eloquent presentations. I want to interject with counter-thinking and ask you about suspension of disbelief. Can you talk about any experience where you had to go through the suspension of disbelief? For instance, Jon, you were talking about the steps of arriving at a form through a mathematical thought process, the stage of challenge; then when that challenge collapses, what happens?

JK I think actually this happens on a very regular basis. There is a psychological element to doing mathematical research: you want to solve a problem or get to a certain end point. You go a certain distance and you reach a point where there seems to be an obstacle and it seems to be insurmountable – that can be very discouraging – but you then have to suspend your disbelief about that and ask: what if I could get round it, what would the world look like on the other side? You have got to force yourself to look at this obstacle from a different direction and perhaps train yourself to think that it could be solved. Almost the act of doing that allows you to solve the problem. It is almost like you have got to persuade yourself that disbelief can be suspended before you can then circumvent the problem or cement it in some way. It is a key part of, in my view, doing research, forcing yourself to constantly say: I think the world is this way but I am going to suspend that and persuade myself that it might be some other way that I do not understand, and then try to reconstruct from that point of view.

FÜ If we used your climbing the mountain analogy, is it like building an additional bridge to take a leap, to come back to where you were?

JK Yes. I am not a mountaineer but I have watched films about it. You get to a certain overhang and you think: no one has managed to get over this before and therefore I cannot do it, but the people who do get around that overhang are the ones who persuade themselves that it is possible, that the disbelief is actually just psychological.

16. Maria Montessori (1870–1952) was an Italian physician and educator, noted for her philosophy of education.

I think a lot of the way that one trains one's students is, when they come to this obstacle to just think: well now, enjoy, play with it, try to find any way around it, you can do it. It happens occasionally [but] a lot of the times you fail and a lot of mathematical research is constant failure. The resilience that one needs to then pick oneself up is very important, but if you train yourself to be resilient you find that some obstacles can be semantic.

FÜ Thank you. Maya, thank you for giving the examples of what could be in a volume of *Collapse*. I believe there is this invisible balance as an editor; even though those subjects feel so disparate and diverge they actually still balance each other. Do you have any experiences where it falls off? Is *Collapse* about that idea of never having that balance?

MBK No, it certainly is a balance and when I talk about it as being for me a kind of a sculptural experience that is very much what I mean. If I have any skill at all at being an editor it is in stepping back, looking at how all these pieces fit together, and thinking: it needs something more here or we need to change the order of these things, until you reach some kind of satisfactory state where you think this is a real thing, this gels together and I can commit to it.

I was just thinking about suspending disbelief. In philosophy it is also the rule rather than the exception because all of the great philosophers are total maniacs. To think that you can expound a system that is going to explain everything in the universe ever you obviously have to be crazy. Obviously if you start doing that, thinking that it is going to be the one true system, then you know you have really lost your mind. Equally you have lost your mind if you think you are going to start reading philosophy and pick at it and tell the philosopher why they are wrong. Because then you do not get anywhere if you just read page one and say: I object to this. What you have to do is enter into these grand, gargantuan fictions that philosophers create and see how the world looks, see what it does to the world, because every philosopher is really more than anything a style of seeing the world, a style of thinking, a style of experiencing. It is always to do with suspending disbelief and listening to this story and every now and again being so captured by it that you take something of that style away with you and you live it every day.

FÜ That is very beautiful, thank you. Emma for you, you could imagine it curatorially, through making an exhibition or maybe sometimes discovering an artist; perhaps in the beginning you do not think their work is strong, [so] you have to dive deeper. How would you like to take this question? It is very open.

ER At school, when you are making your decisions about key subjects, I was hedging between philosophy and art. I chose art professionally, yet I have continued reading philosophy as a hobby but [as a] completely shameless amateur. [I] kind of just enjoy it and dive into it because there are ethical ways of understanding the world that I get a lot out of.

Similarly, with artists or projects I choose, usually there is some resonance of values. It sounds an unusual thing to say, and it always startles students when they ask about my motivation: what has led to your projects? why are your projects very zeitgeist? I say, oh, it is *anger*: I am attuned to my body in that way; when something is really frustrating me in, say, public discourse, or there is something not being said, or there is some frustrating rigidity about how things are being looked at, I am motivated to address that. When I had been in Oxford for not too long (by way of example) some of the conversations around race and gender seemed to me rather assumptive and frustrating. This was in 2015. So, I invited Lubaina Himid to have her first institutional exhibition as an artist who had done brilliant stuff since the 1980s, opening up and exploring questions on race and gender through a very subjective yet very generous open lens.

For me, exhibitions always start from a question or a point of frustration. I do not mean that in a negative way, but exhibitions are part of a public conversation, they are a chance to debate things and explore ideas in very diverse ways. So, as part of contributing to that public conversation, thinking about which artists would resonate or be able to say something in a stronger way than I could, and take people on the journey with them. I think you should be changed by what you are working on and be open to learning, such that you come away somewhat different too. Exhibitions take such commitment. Making exhibitions is harder to do than you might imagine. The logistics are completely up against you all the time, and that is the bit no one is interested in, so you have to just suck it up and not even talk about those bits. It is a kind of optimism over reality to commission art or get exhibitions to happen anyway. To my mind, to have that level of commitment and passion in there (the same with choosing to lead the Foundling Museum) you must be so in love with something, or see the potential, or the beautiful DNA that is within it, to make that commitment and leap. You do not know what it is but sometimes there is something intuitive or something that speaks to you, the bit of grit in your sense of the contemporary space in the moment and what you might put into that (with others) to nudge that along.

FÜ Amazing, thank you, what about you Conrad?

CS I have always used a sort of device, a sort of cloak of those sculptures which is a bit of a disingenuous device or tactic. I remember one time I was installing something in a window and someone rushing past stopped and said: what is it? I began to explain that it was an artwork and before I could finish, he was like: oh, and then just walked off. He was really interested in it before, because he had to deduce the function from the form and there was a sense that this thing had some kind of useful purpose or some kind of rational intent. It was very useful, this reaction, because it made me realise how powerful it is to give these things this cloak of the rational or of a scientific or a mathematical model.

With all these works, and it is why it is so nice to see them in this building and in this context, there is this energy, where they oscillate between successful models or something [else]. They create intrigue and questions; all these augmented minds within the institute are seeing something different within each one. I have often used the suspension of disbelief in that way, where I am constantly trying [to] give the pieces the authority of the rational. I do not want to be whimsical or 'Heath Robinson'[17]-like – it is one of the reasons I moved away from wooden structures in my early machines, as I wanted them to have the authority of the real, the authority of the machine to lead people down a kind of garden path where you assume this thing had a logical function. Actually, in fact, it is irrational and functionless but through those devices it asks many more questions than it answers and hopefully has poetry. I think the proof of a good artwork is that everyone has a different opinion about it and that breadth of interpretation. If there is a consensus around the object, it is probably not a very good artwork, but when everyone sees something different within it, then you really think you are getting somewhere. You lose control, it is no longer your possession in terms of the meaning, and that is when I get really excited, if I lose control of that meaning.

FÜ Thank you. It is a brilliant segue to our last question before we open to the floor. We ask our speakers to pick a work from the exhibition that resonates with them. Maybe we could start with you, Jon.

JK I chose this one [*Bicameral*] of which there are two versions by the back door of this building. You will see it, perhaps, as you leave. What I like about this is the fact that it is a tree-like structure and that each branch breaks up into smaller branches, so you can imagine this going on forever. It is an infinite structure in a finite volume, so it is almost a good example of Ruth's work that you were speaking about earlier.

17. Heath Robinson (1872–1944) was an English cartoonist, illustrator, and artist noted for his drawings of elaborate machines.

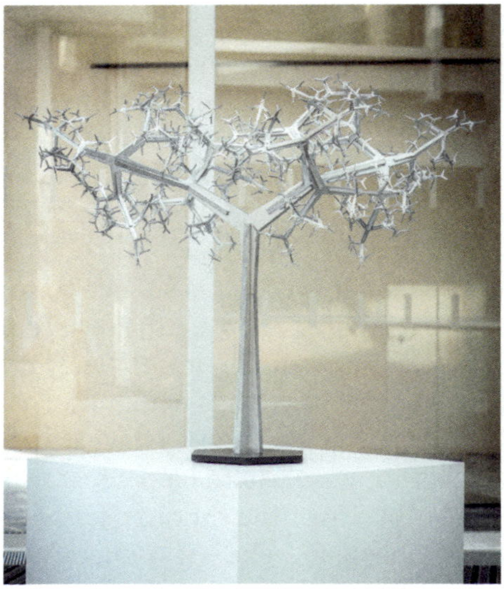

Bicameral Study, 2019

It is in my language a fractal structure: you zoom into a tiny bit of this sculpture and what you see is an almost perfect reconstruction of the whole thing. That for me is tremendously important, but the thing I love about this is [what] I mentioned earlier, this understanding that we have of the hills and valleys of the Riemann zeta function. The formula that generates those hills and valleys has exactly this tree-like structure and we never knew that before, up until about ten years ago. Now we cannot look at that formula and not see something like this, and so every time I walk past this, it reminds me of the formula. Also, I think it captures beautifully the beauty of the formula on a daily basis: that is amazing.

FÜ Emma, we have already seen a glimpse of the piece that you have selected [*Fracture*].

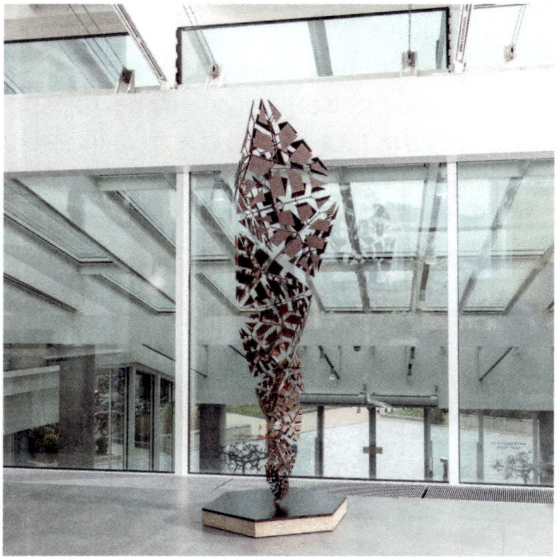

Fracture, 2018

ER Having seen for a very short moment the works of Ruth Asawa [via] images, you may see that this, to my eye, has some connection. Why I am particularly taken with this work [is] to me it always seems like it is slightly moving, yet I know that it is not. It has that lovely dissonance – you have the sense that something is growing and moving but your rational brain knows that it is still and fixed. In terms of the relation to the body as well, it is very person sized, which I really like.

Lao Tzu[18] was a big influence on Joseph Albers, when he was teaching the students about negative space and the importance of understanding the void. To my mind [it is] one of the best ways to understand Asawa's wire works, and perhaps *Fracture*. As many of you will know, in Eastern philosophy, the void is *the* essence and the most important aspect of life, the energy of things. This is a beautiful articulation by Lao Tzu:

> Thirty spokes meet at the hub but the void within them creates the essence of the wheel. Clay forms the pot, but the void within creates the essence of the pot.

Essentially the material aspect creates the unity of the form but it is the immaterial aspect (that we cannot see) that science has now revealed. There is plenty there, but we think of it as void or immaterial because we cannot see it. It is this immaterial void that is the essence of life. In Eastern thought (called 'Ma' in Japan) it is a much more familiar concept. It is that tandem of material and void that speaks to me in *Fracture* and the work of Ruth Asawa.

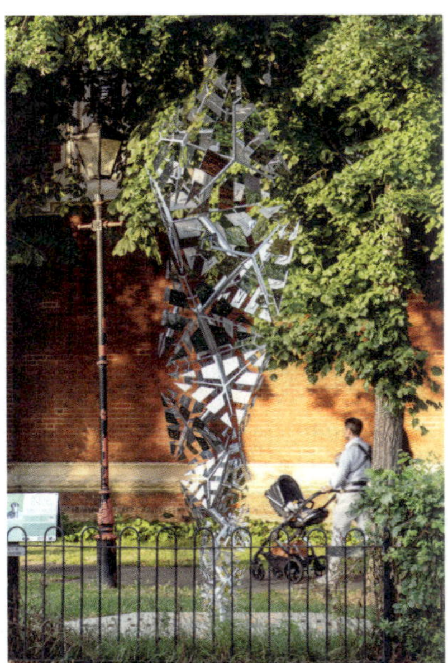

Fracture (R45S33) - Enwrought Light, 2021

18. Lao Tzu was an ancient Chinese philosopher and author of the *Tao Te Ching*, the foundational text of Taoism.

CS Thank you, that was really beautiful. I thought maybe it would [be] a good moment to show the Yeats Memorial, because it is linked to this piece. It also links I think to these ideas of 'we are because we are perceived'. This is a piece that follows on from the piece that Emma kindly just talked about and it has got the same geometry; it is called a *Fracture.*

I was asked to make a memorial to the poet Yeats and so it has got this sense of a widening gyre. Instead of it having these rusty leaves all over the surface which are all unique and suspended almost like a light bulb that has shattered before it all falls to the ground; there is this sort of ghostlike field.

In terms of what you were talking about, Jon, about the idea of quantum mechanics, [Schrödinger's cat][19] and this philosophical question, with this work you as you walk around it or as the sun shifts, the piece only exists because of the reflection in the surfaces. You see yourself fragmented within it, you see the sky, the seasons, the sunset, the sunrise; it is constantly shifting but it only exists because it reflects things around it. It owes its identity only [to] being perceived, by light or being hit by light. Without that it sort of disappears.

Again, with the Ruth Asawa [comparison], it has got hopefully that really nice thing where you really want to walk around [it] and see between it the interference patterns of the back and front.

FÜ That is very beautiful, thank you Conrad. Maya, maybe we could come to you.

MBK So, this piece, *Paradigm*; I think what intrigues me about it is this structure whereby it obviously is a sequence. It has a kind of upward aspirational élan. Some kind of developmental process seems to be going on and I can tell there is something mathematical happening, but I cannot make out the exact nature of the process, how each shape develops, one from the other. It has a definite direction but it is somewhat unpredictable. In the title *Paradigm* I see a reference to Thomas Kuhn's[20] famous theory of the scientific paradigm shifts whereby science is ultimately conservative. It wants to retain conventions but every so often there is this kind of breaking point where it has to give way, and a whole new kind of paradigm takes over.

This makes me then think about something that has often occurred to me when trying to bring together philosophy, art, and

19. Schrödinger stated that if you place a cat and something that could kill the cat (a radioactive atom) in a box and sealed it, you would not know if the cat was dead or alive until you opened the box, so that until the box was opened, the cat was (in a sense) both dead and alive.

20. Thomas Kuhn (1922–1996) was an American historian and philosopher of science who introduced the term 'paradigm shift'.

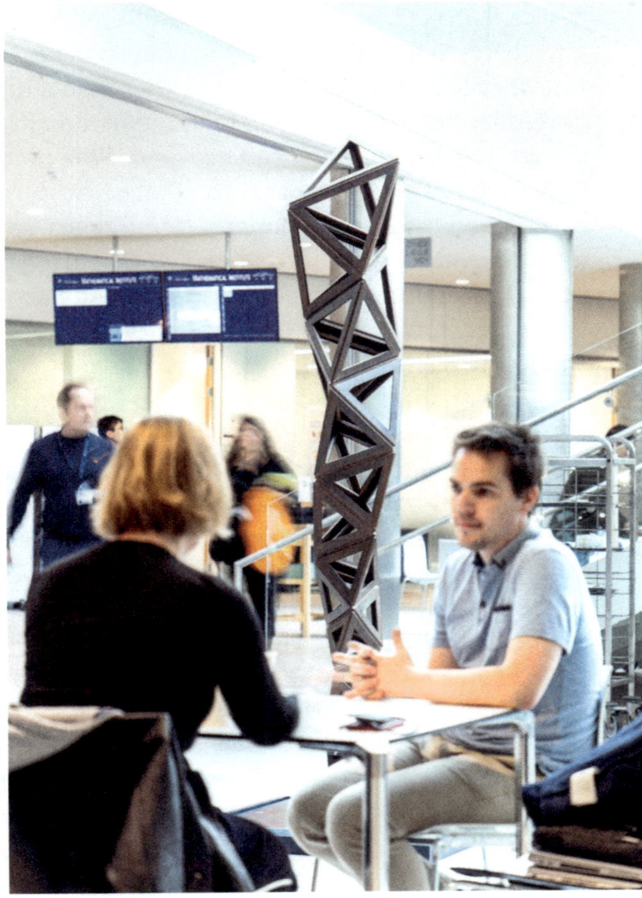

Paradigm Slender (Structural), 2016

science; which does not always work. These practices have their own different relation to their own history. If you take science, there are these paradigm shifts but the way some people misunderstand this is that science does not just throw away everything that went before and start again. What happens is that what was there before usually becomes integrated into a new model as a special case, as happened with Newtonian and relativistic physics. Philosophy is a weird one. It takes you a long time to establish yourself in the history of philosophy; probably I would say at least ten years to really start feeling confident and at home. During that period what happens is all the deep thoughts that you thought you had you find out that someone else has had them, and they all have names, and they are all very well-established questions in philosophy.

These are established landmarks, discoveries, and even innovations in philosophy. But at the same time, it has a peculiar relation to its own history; it continues to inhabit its entire history in a way that other disciplines do not. There is no invalidation, there is no real superseding of one philosophy by another, so you can always go back to Plato and indeed we always do. There is a kind of circulation within the history of philosophy which to an outsider makes it look like a massive waste of time because you never get anywhere. What is important in philosophy is that we continually return to the same problems and we refine and restate them in new, different ways. In art, if you take the paradigm of contemporary art, if anything it means you can use any of the practices and materials that have been used in art throughout history. There are still some established historical moments. Modern art as such is now over in a sense, and for some artists the worst thing you can say is that it has been done before. The art world sees itself as this kind of collective inquiry in which certain historical moments are established and you cannot go back, and by doing that it kind of disconnects itself from a kind of art that would be just to do with the beautiful, just to do with personal development. All of that goes into the bucket of art therapy. I am not saying this is the paradigm I agree with, but I think this is the paradigm of contemporary art. That is a historical thing. But also remember that history is being reinvented. It was not until the '60s that Marcel Duchamp[21] was rediscovered and crowned as the great forerunner of everything that would come after. There is always this delving back into history and reinvention.

Mathematics it seems to me is a particularly interesting one. Is mathematics about discovering things or is it about inventing things? You certainly do not get all of mathematics at once, it occurs within history but it is supposedly a self-consistent structure of truths which are in some sense eternal, always true. What exactly do mathematicians do? I think Jon has enlightened us a bit already. It cannot simply be a case of logically deducing one mathematical truth from another, otherwise you could do that very quickly. Why does it take so long and what exactly are mathematicians doing? Usually in philosophy of mathematics people do not talk about this – they talk about what is called Platonism: what are mathematical entities, do they really exist, do they exist in some other realm, why are they applicable to things in the world? I think what is really interesting about philosophy of mathematics is this question of its historicity, not what its objects are but how it happens, because, ultimately, it is the question of how something new is possible. In the European tradition of philosophy this has been I think far more prominent.

In the interest of brevity, I just wanted to introduce you to two very interesting French philosophers, both of whose lives were tragically cut short by the war. Jean Cavaillès[22] in fact is a national hero in France, but both [he and Albert Lautman[23] were] in their own way. They were friends but their positions were

21. Henri Marcel Duchamp (1887–1968) was a French painter, sculptor, chess player and writer, whose work is associated with Cubism and the development of Conceptual Art.

22. Jean Cavailles (1903–1944) was a French philosopher and logician who specialised in the philosophy of mathematics and science.

23. Albert Lautman (1908–1944) was a French philosopher of mathematics.

very different. [They] tried to address this question of what happens within the practice of mathematics; how can it be at once contingent and necessary, how a new mathematical truth is produced. That is really what fascinates me and I think Conrad's works, in the way they cross over art, science, and philosophy, have made me think about these kinds of problems, which do not belong to any one of these disciplines, which in a sense they share.

FÜ That is very beautiful, thank you so much Maya. I think this is the moment for a magical turn, so if we could turn off the lights.

CS We have had a tradition here where I have brought some kind of shadow maker or light maker. This is [the] last I have made recently of the *Slow Arc* series, which started off with a very simple mesh cage and [has] evolved into these double layer, double lights, and different experimentations. Each time I make one of these cages it is about a simple premise of moving a light from one corner of the cage in a sinusoidal speed profile to the other. This one has a single light and a double skin with a very particular mathematical pattern that repeats on both surfaces. Like with Asawa, the piece has a kind of interference pattern through the skin and as you walk around it you get the moiré effect.

One of the big questions with these works is: are you able to get back to the object from just seeing the shadows? The last one I showed in this room had a rhombic mesh and there was a triple cage; some of you may have seen that. The rhombic mesh has probably 80% of the material removed through each layer, so you are getting more than 50% of the light penetrating through the layers, the Russian doll of the cage. The perception of that is you see the structural shadows on the wall and they are perceived as shadows.

What is interesting with *this* work is that the amount of light that is able to escape through the skin is less than 50% and suddenly the shadows are no longer there. All you see are these dots or these particles, and so it suddenly becomes more starry, more like moving particles of light and that is a perception thing. The critical point of moving past 50% means the way you perceive this thing shifts but then what you see through this interference of these layers is that a particle appears, then drifts across the sky, and then disappears, and then reappears. You are getting these moments of disappearance, reappearance, and interference, and these perspectives. I think again this for me is really analogous to us observing the universe and trying to figure [it] out. As Jon mentioned, being in the dark room is a perfect metaphor here. We are trying to turn on the lights, we are all trying to see beyond our perception envelope. As artists we are trying to see that through the endless repetition of an idea on canvas or in sculpture, of carving or adding. With mathematicians the endless experimentation, the refinement.

All those descriptions that Jon made, all of Asawa's work, and [for] philosophers it is all about trying to see ingeniously beyond our perception envelope and to work out the truth of why we are here, what we are in, and how we inhabit this strange and extraordinary sort of environment.

I work as an artist in very much the same way as maybe a mathematician and maybe a scientist and each time I make one of these cages I just change one thing and I do not know what the result will be. Sometimes they are not that successful and sometimes it is extraordinary but it is often just the simple repetition of something simple [that] leads to something really quite breathtakingly complex. That is always why that eureka moment is so exciting when you turn the piece on for the first time and see the effect, which is very different from what you expected.

FÜ Thank you so much Conrad, and thank you so much for coming today. Thank you, Emma, Maya, and Jon, for your immaculate contributions. Thank you to the Mathematical Institute and Dyrol for making this happen.

CS Fatos has been extraordinary to work with on this wonderful thing, and of course the Maths Institute has been wonderful to host this.

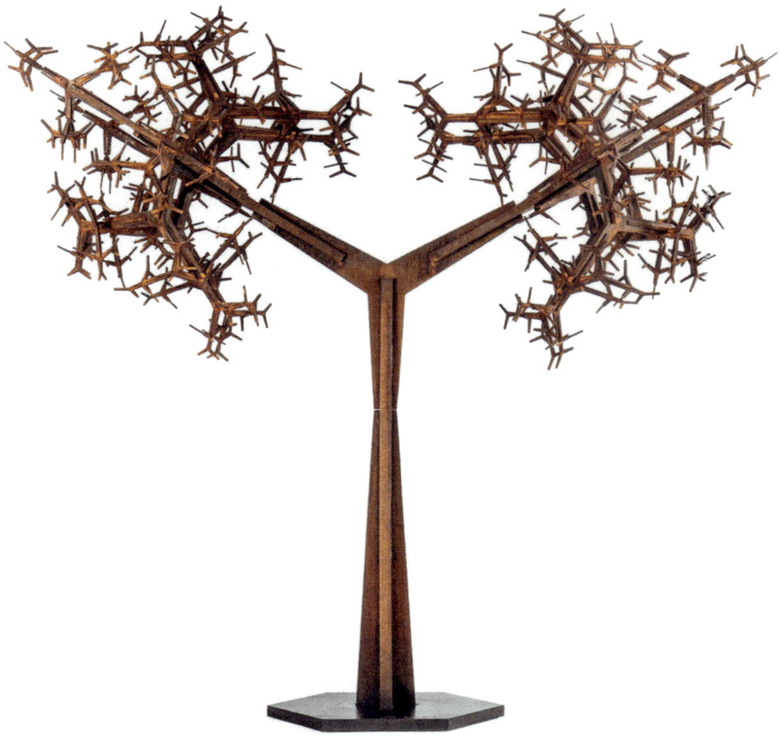

Bicamerals

The term *Bicameral* was coined by Julian Jaynes in his book 'The Origin of Consciousness in the Breakdown of the Bicameral Mind', which theorised that consciousness evolved out of the use of language, particularly metaphors. The sculpture is a progression of bifurcating and trifurcating elements that fan out from a stem, which loosely forms two hemispheres or sides. *Bicameral*'s creation borrows its technique from traditional Japanese joinery, formed with forked connections, without the need for welding or screws. The work has been realised at different scales, from 8m down to 60 cm, and in aluminium, anodised aluminium, and weathered steel. A 8.5m *Bicameral* rendered in blue anodised aluminium is on permanent display in Chelsea Barracks, London

Primes, Trees, Glasses, Freezing – the Mathematics

Jon Keating, Mathematical Institute, University of Oxford

The primes are the whole numbers that are divisible only by themselves and one: they cannot be expressed by multiplying together smaller whole numbers. The list of primes starts 2, 3, 5, 7, 11, 13, 17, 19... and is infinitely long. The primes are extremely important because all of the other whole numbers can be produced by multiplying them together; for example: $6 = 2 \times 3$, $36 = 3 \times 3 \times 4$, etc. They are like the atoms of arithmetic. There are still many elementary things we do not know about them, and they remain a major focus of mathematical research.

One way to explore properties of the primes is by studying the following object:

$$\zeta(s) = \frac{1}{(1-\frac{1}{2^s})(1-\frac{1}{3^s})(1-\frac{1}{5^s})(1-\frac{1}{7^s})\cdots}$$

This is called the **Riemann zeta function**. The right-hand side involves a product over all of the prime numbers, and $s = x + i\,y$, where x and y are numbers and where i denotes the square root of -1. One can think of this as providing an algorithm that produces an output number from two input numbers, x and y.

Many fundamental properties of the primes relate to the pairs of input numbers that produce an output number 0. These particular pairs are called the **Riemann zeros**. The Riemann zeros either have $y = 0$ and x equal to a negative even whole number, e.g. $x = -2, -4, -6$ etc. (these are known as the **trivial zeros**), or they have y not equal to zero and x between 0 and 1 (known as the **nontrivial zeros**).

The nontrivial zeros are extremely important in mathematics. They are the subject of one of the most famous unsolved mathematical problems, the Riemann Hypothesis, which asserts that they all lie on the vertical line $x = 1/2$ in the x-y plane. The function is plotted along this line in the graph below; the zeros are the points where the curve crosses the x-axis. There are infinitely many nontrivial zeros and we know the Riemann Hypothesis is satisfied by the first 12 363 153 437 138 of them, but it remains a deep mystery how to prove it in general.

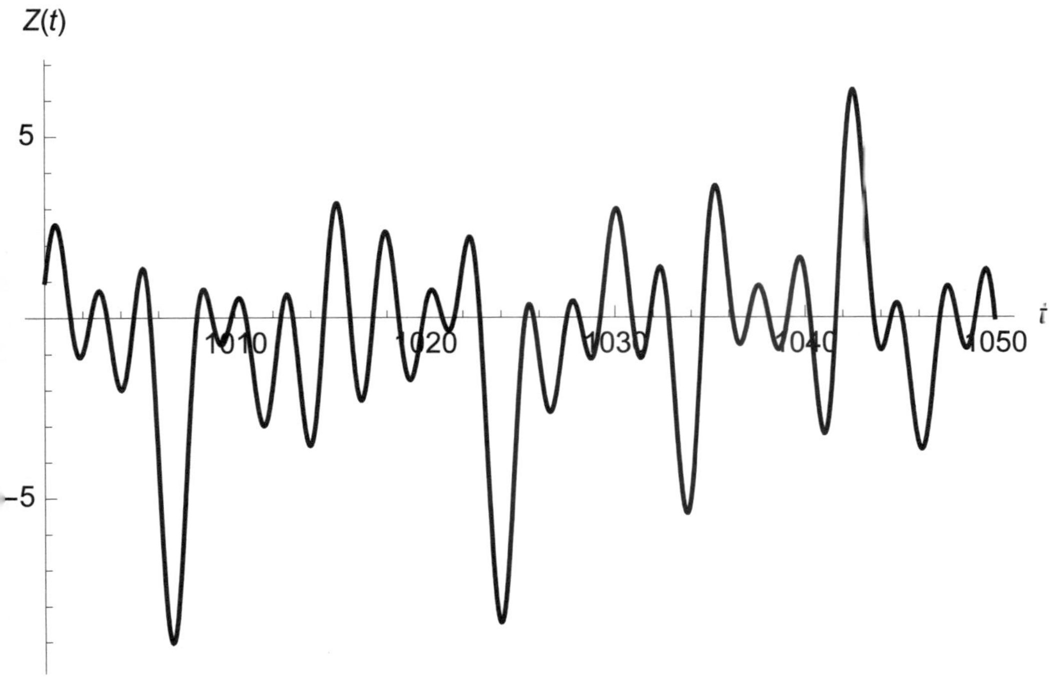

Representation of the Riemann zeta function on the line $x = 1/2$

Another deep mystery is how big the Riemann zeta function gets on the line $x = 1/2$; that is, how far from the horizonal axis can the curve in the graph find itself? Resolving this has been a key challenge for over a century. It was conjectured in 1908 that its values do not get too big. This is known as the **Lindelöf Hypothesis**.

Until recently, we had no precise understanding of how big the values of the zeta function on the line $x = 1/2$ values can get. However, over the past ten years, there has been remarkable progress on developing a precise understanding, motivated by ideas from completely unexpected directions.

The first insight was that determining how big the values of the zeta function can get bears a highly surprising similarity to a problem in physics concerning the freezing of glass-like materials. We have a relatively clear understanding of how simple materials freeze – that is, how they go from a liquid to a solid phase as temperature is lowered through a critical value (the freezing point) – based on how their internal energy depends on the configuration of the atoms or molecules from which they are constituted. In glass-like materials, however, the internal energy depends highly sensitively on the configuration, such that it behaves in a seemingly random way. The problem of freezing is related to how small this energy can get, and when one expresses this mathematically it bears an uncanny resemblance to the problem of understanding how far from the horizonal axis the curve in the graph gets. Put in a nutshell, the values of the zeta function seem to freeze like glassy materials!

Based on this clue, we have, over the past ten years, understood much more clearly the big values of the zeta function by using ideas from physics to motivate new perspectives on the mathematics. In particular, ideas from physics have led us to identify a remarkable feature of the product over primes in the formula for the zeta function that had eluded us for over 150 years. We now understand that the primes can be collected into groups. If we take two different positions on the vertical line $x = 1/2$ (that is, two different heights), some groups give similar contributions to the product and others give wildly different contributions.
The closer the heights, the more groups there are giving similar contributions. The positions can be pictured as being connected to a tree-like structure, on which some points have many branches in common in the way they are connected to the root, and others have fewer. That is, the way the product over primes generates the values of the zeta function resembles Conrad Shawcross's magnificent *Bicameral* sculptures. Knowing this has led to astonishing progress in understanding how big the values of the zeta function can get.

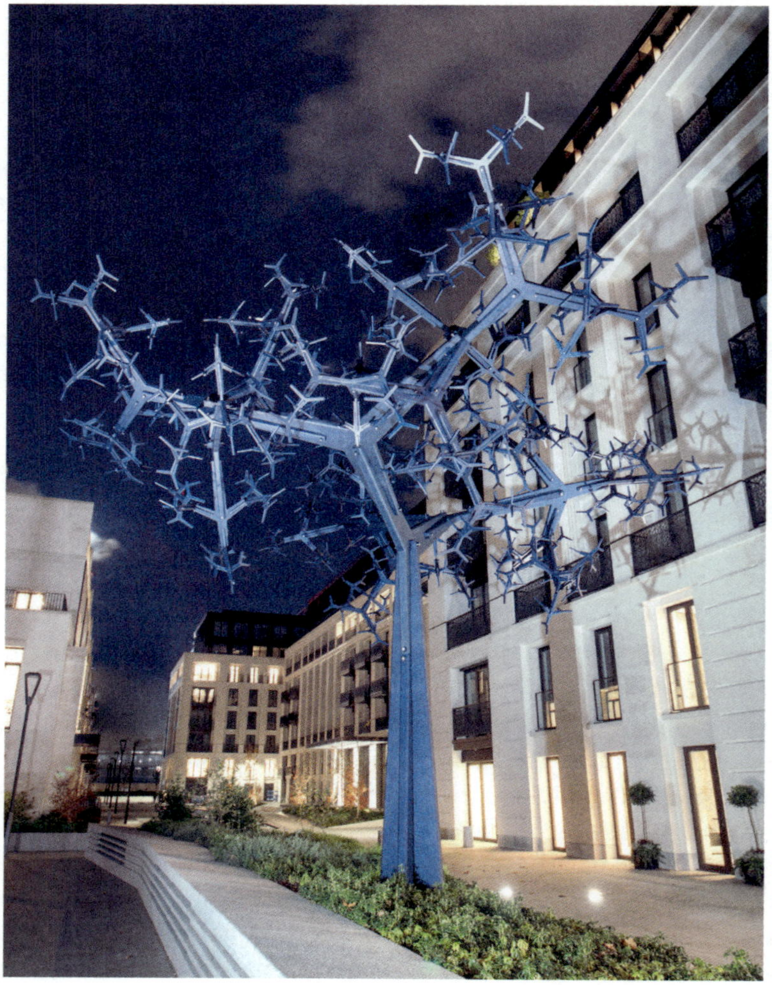

Bicameral Study, 2019

Primes, Trees, Glasses, Freezing – the Image

Chrystal Cherniwchan, School of Arts and Media, University of Salford

When creating visual artworks in response to a subject that is about specificity and where the language is constructed with clear definitions, in this case mathematics, the intangible and mysterious phenomena in human and more than human existence(s) seems to be a shared fascination for both the artist and mathematician. There may be a point, of course, when a shared curiosity diverges and we find that the mathematician is motivated by the conjecture and proof, whereas the artist can be driven to create visual questions that engage audiences to form their own conjectures.

The journal article titled 'Freezing Transition, Characteristic Polynomials of Random Matrices, and the Riemann zeta function', by Yan V. Fyodorov, Ghaith A Hiary, and Jonathan P. Keating, influenced, in 2013, a series of photographs titled 'Freezing'. The aim was to highlight the slippery nature of unknown territories, horizon lines and the oscillation that occurs on those lines; and a recurring theme of interest, what happens outside the photographic frame, the multiplicity within our multidimensional spaces/narratives, and how that links with the 'deep mystery of how big the Riemann zeta function gets on the line $x = 1/2$'; that is, how far from the horizontal axis can the curve in Jon Keating's contribution find itself?

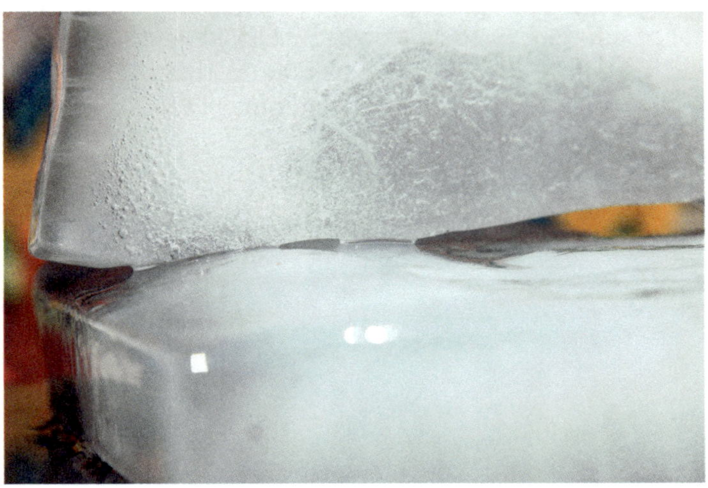

Chrystal Cherniwchan, *Untitled* [Photograph], From the series *Freezing*, 2013

Blocks of ice were made and photographed in a way to suggest a landscape where the field of view extends beyond vision, and the use of synthetic colour was included to imply northern lights or hypothermic confusion. Showing a plurality of events was important, including scientific moments such as light waves bouncing off surfaces and their reflections, freezing, melting, and the threshold in between, and trapped air bubbles like fossilised breath. An additional layer in the process when recording these events, is how we read the photograph and interpret the image, and this is often done with metaphorical language that circles around subjectivity; possibly a different reading from the scientific community.

Photographs inherently flatten dimensions expressing a collision of ideas. French philosopher Gilles Deleuze and French psychoanalyst and political activist Félix Guattari use the rhizome as a concept representing non-hierarchical structures and/or nonlinear ones.[1] Mycelium networks and the collaborative workings between multiple life forms are another example, and how they connect to photography can be found in both the materials of photographs and the equipment needed to make them, together with the visual representation of time in a fraction of a moment.

This visually links to what we see when looking at Conrad Shawcross's *Bicameral* sculptures, the systems that coexist beyond the sculpture itself, and beneath the base of the structure, such as imaginary roots. In addition to the sculpture there is also the research and dialogue with scientists that helped shape the work. Anthropologist Anna Lowenhaupt Tsing writes, 'landscapes are not backdrops for historical action: they are themselves active. Watching landscapes in formation shows humans joining other living beings in shaping worlds'.[2] An echo of our collaborative working.

Regardless of the language or tools that one uses to question the curiosities that coexist amongst us, the shared aim is to provide further knowledge and understanding, albeit the gradient of 'truth' that the mathematician and artist seek differ, and those differences could also be questioned and defined along its own horizontal axis, but we would agree that the possible infinite distances are an exciting place to find ourselves.

1. Deleuze, G. & Guattari, F. (2002). *A thousand plateaus: capitalism and schizophrenia* (B. Massumi, Trans.). 'Continuum'.
2. Tsing, A.L. (2015). 'The Mushroom at the End of the World'. Princeton University Press, p. 152.

Maya B. Kronic

The title of Conrad Shawcross's sculpture *Paradigm* invokes philosopher of science Thomas Kuhn's famous theory of paradigm shifts: a given field of science, Kuhn suggests, tends to be conservative and resistant to change, up to a certain point, and then the consensus will suddenly shift toward a fundamental new structure, and established scientific facts will be viewed in a different way.

Shawcross's sculpture itself presents a sequence, a kind of upward aspiration or élan, some kind of developmental process. And, as in all of Shawcross's works, the viewer can tell that there is a mathematical basis to this sequence. But it is not easy to make out the exact nature of the process, which has a definite direction and yet a certain precariousness.

This form brings to mind something that has occurred to me often in my role as an editor: when trying to bring together between the covers of a book the different practices of philosophy, art, and science, one thing that becomes clear is that they have a very different relation to the sequences that make up their respective histories.

There are indeed paradigm shifts in science, but it would be a misreading of Kuhn to imagine that they involve throwing everything out and starting anew. Rather, an existing model is usually integrated into a new model as a special, local case – for instance, Newtonian physics was not invalidated, but encompassed within special and then general relativity.

How about philosophy? It takes a long time to establish yourself confidently in this field precisely because it is so tied up in its own history. Along the way, unless you are an exceptionally original thinker, you will learn that your treasured deep thoughts have all been thought before, and have well-established names and terminologies attached to them. Yet that does not mean they have been solved. There are landmarks, discoveries, and innovations in philosophy, but it maintains a peculiar relation to its own history, continuing to inhabit it in a way that other disciplines do not. You can (and we do) always go back to Plato, there is no real 'invalidation' or 'superseding' of previous philosophies. Philosophy circulates within its own history in a way which, to outsiders, can look like a patent waste of time. But ultimately what is important here is not finding *solutions*, but the refining, recasting, and revisiting of cardinal philosophical *problems* that never really go away.

As for contemporary art, it certainly invites an artist to use any and every technique and approach available from the history of art and adjacent fields. However, it also softly whispers that there are paradigms and watershed historical moments which, unless you are willing to risk ignominy, cannot be ignored. The strivings of romantic aesthetics and the disciplinary distillations of modern art, for instance, are passé unless you are using them as a reference in some kind of critical operation. Generally detaching itself from any idea of art as personal expression, contemporary art likes to present itself as a kind of collective enquiry into the now, albeit one whose knowledge criteria are rather vague. It is highly critically conscious of its history, but it bears remarking that this history itself is continually being reconstructed and reconfigured – it was only really in the 1960s, for example, that Marcel Duchamp was 'rediscovered' as a crucial, unavoidable figure in the prehistory of contemporary art.

Mathematics is a particularly interesting case. Is mathematics about discovery or invention? We do not get all of mathematics at once, it develops within history, and yet what it reveals is supposedly a self-consistent set of adamantine, eternal truths. So, what do mathematicians do exactly? Just logically deduce one mathematical truth from another? If so, why does it take so long?

Philosophy of mathematics is often seen as involving the question of what is called 'Platonism': Do the objects of mathematics 'really exist' somewhere, or are they convenient fictions, useful conventions? But the most profound philosophers of mathematics, I think, have instead posed the question of the *historicity* of the discipline – which, ultimately, is the question of whether something new is possible, and how.

European philosophy of mathematics has generally taken more of an interest in history than the Anglo-American tradition, probably owing to the legacy of one philosopher: Hegel.[1] Broadly speaking, throughout its history Western philosophy aspired to respond to questions about Being: What can we say about what is, always has been, and will be? What are the primary characteristics of any and all beings? But in the late 19th century, Hegel, himself living through some momentous historical events, changed the form of this question. To put it simply, he decided that Being – the question of what is and how it is – is not of primary interest for philosophy. What is more important is to look at how the Being of the world relates to itself, and reshapes itself to produce new forms. It relates, of course, through thought and consciousness, or what Hegel calls Spirit or *Geist*. And the crucial thing Hegel tells us is that Being's relation to itself changes across history; there is a sequence of 'shapes of thought', forms of *Geist*, in which Being relates to itself in new ways.

Trains of thought, Hegel argues, tend to run to incompatible extremes, which are then synthesised under a new paradigm. This is how we get the famous Hegelian dialectic – the story of Spirit shapeshifting across history,

1. Georg Wilhelm Friedrich Hegel (1770–1831) was one of the most prominent figures in German Idealism, and a massively influential figure in modern philosophy.

the story of how Being relates to itself in new ways which then end up producing new incompatibilities that must be resolved in turn. For Hegel, this process, in fact, *is* history (which has an end – the point at which all incompatibilities are resolved, and pure Spirit reigns). Without it, there would be no true *becoming*, but only the same furniture of Being rearranged in different ways. History is where something happens and, according to Hegel, a new kind of philosophy is needed to analyse it: dialectics.[2]

Against that background, I would like to present a brief portrait of two French philosophers of mathematics, both of whom were thoroughly steeped in this lineage of philosophical thought as well as in the most recent mathematical developments of their time.

Jean Cavaillès (1903–1944) wanted above all to talk about the experience of *making* mathematics; he explored mathematical *experience* (a word which, in French, means 'experience' but also 'experiment') in the hope of understanding how the truths of mathematics also participate in a *becoming*: a dynamic, historical process which does not have a predestined outcome. Cavaillès insists that there is true novelty in mathematical history: it is not just that someone has realised what was already obvious, secretly contained in the premises of the work of the previous generation of mathematicians. Mathematical experience is not just logical deduction, it is a historical (and perhaps dialectical) building process. Mathematical *objects*, Cavaillès argues, have to be thought alongside the *gestures* that are used to build them – the different kinds of mental manipulation with which mathematicians handle the objects and structures handed down to them, and shape them to produce new objects, new structures, which can then be used as building materials by the next generation.

Two of the gestures Cavaillès explores are *thematisation* and *adjunction*. *Thematisation* happens when operations upon objects become objects in their turn. In group theory, for example, transformations or 'morphisms' between objects become a new type of mathematical object that can in turn be worked with, and a whole new set of operations can then be performed upon them. *Adjunction* takes place when you have an established field of objects defined by rules which enable you to consistently carry out certain operations upon them – and then you add a new object which, according to those rules, is 'impossible'. The most startling example of this is the complex plane, which results from the introduction into the number system of an 'impossible' number, the square root of minus 2. The invention (or discovery?) of the complex plane produced a peculiar

2. This is the idea that Marx would turn upside-down (or, according to him, right way up), developing a *materialist dialectics* where, instead of the struggle between thoughts and their eventual synthesis, the sequential incompatibilities and resolutions of historical becoming are driven by class struggle.

new mathematical structure, and made possible a new set of operations which, quite counterintuitively, soon turned out to be applicable to certain problems in physics.

In short, Cavaillès sought to produce a philosophy that would address mathematics as a practice involving the forming of thoughts, and began to compile a kind of toolbox of experimental gestures by which they are so formed.

Cavaillès was good friends with Albert Lautman (1908–1944); both philosophers died young, executed by the Nazis during the Second World War (Cavaillès in particular is celebrated in France as a hero of the Resistance). The two of them had different philosophical positions, but they did agree on the important questions to be asked in the philosophy of mathematics.

The background to Lautman's thinking lies in the early 20th-century work of the great German philosopher Martin Heidegger,[3] who, in his own idiosyncratic way, took up Hegel's idea that philosophy should not address eternal truths, but the 'history of Being'. Unlike Hegel, however, Heidegger did not see world history triumphantly culminating in the modern European nation state. According to Heidegger, what is characteristic of modernity is the *forgetting of the question of Being*. We have all sorts of discourses about *beings* – physical beings, chemical beings, atomic beings, biological beings, and so on. We even have a great overarching discourse about how we can analyse *all* beings and control them – the discourse of technology and cybernetics. But all of this, for Heidegger, is synonymous with a forgetting of the real question of Being, which involves something like the wonder or poetry of the fact that there is anything at all, and a contemplation of the endless dynamic prodigality of the world around us.

In the face of this dangerous forgetting, Heidegger calls upon us to recognise what he calls *ontological difference*: the difference between *Being*, that mysterious force which gives itself, unfolds, poetises, produces – and *beings*, which are just externalised objects that can be described, controlled, rearranged, and used, but are essentially abstract, produce nothing new, and threaten to make our thinking as inert and uninspired as they are.

As a very early French reader of Heidegger, Lautman had the peculiarly ingenious idea of reading the history of mathematics in terms of this difference between Being and beings. Established mathematical objects are the beings that the working mathematician manipulates: they form impressively consistent structures that are explicitly defined and functional – they 'work'. But they reveal little about what the *Being* of mathematics as *a practice or process* really consists of.

3. Martin Heidegger (1889–1976) was a German philosopher who, through his reinterpretation of philosophy within the framework of the finitude of human 'being-there [*Dasein*]', made fundamental contributions to phenomenology and existentialism.

What underlies the existence of mathematical beings and produces new ones, Lautman says, are *problems* – fundamental, eternally productive problems which are never 'solved'. Certain dualities seem to be 'incarnated' again and again, in more complex forms, by subsequent generations of mathematicians – some examples Lautman gives are continuity and discontinuity, the local and the global, and the extrinsic and the intrinsic. These problematic couplets constitute the inexhaustible Being of mathematics, and the history of the discipline is the story of their endless unfolding.

For Lautman, existing mathematical objects are historical incarnations of these fundamental tensions which, themselves, can never be fully expressed mathematically. Like Heidegger's Being, or Hegel's Spirit, Lautman's prodigal problem-ideas (dualising problems, but also Ideas or Forms in the Platonic sense, in which the beings of the mathematical world 'participate') are mysterious, inexhaustible entities from which mathematical experience continually draws new energy, casting them into new forms but never ever exhausting or fully circumscribing them.

Although their approaches are different, both Cavaillès and Lautman feel the need to speak of mathematics as *experience* and as *experiment*; both of them seek the wellspring of historical novelty within this apparently rational and ordered field that supposedly deals in ideal objects and eternal truths. They seek not the beings of mathematics, but, in Lautman's words, 'that whole historical and contingent becoming that constitutes *the life of mathematics*'.[4]

Which history or histories is Shawcross's *Paradigm* in communication with? Although undoubtedly structurally dependent upon mathematically-grounded principles of engineering, *Paradigm* is not simply an educational illustration of a mathematical concept, nor is it an interpretation representing an artist's free imaginative response to such a concept. His sculptures reach out instead toward underlying *problems* or *gestures* which have a philosophical dimension that exceeds any given mathematical structure, and which by rights belong as much to art as to science. In doing so, they can help orient us toward an understanding of mathematics not as an eternal frozen realm of truths, but as the historical site of an inexhaustible and productive experimentation.

4. For a conversation between the two philosophers, see the 1939 text 'Mathematical Thought', www.urbanomic.com/document/mathematical-thought/.

Conrad Shawcross

I was at the Ruskin in Oxford in the late nineties and, reflecting back now on my time there, I was lucky to arrive at a tremendous moment for the school: in full swing under the leadership of Stephen Farthing, with Brian Catling the head of sculpture.

Brian was an incredibly inspiring, eccentric figure to whom we all aspired. He showed us a liberated way of being that helped us all unwind the tight mechanisms of our more formal educations, whilst being a constant conduit for the proliferation of ideas.

He was a maverick in a pinstripe suit whose work ranged from poetry to performance art. I later understood that we had dyslexia in common, as well as a desire to explore other fields well beyond the traditional boundaries. During the years that he was my tutor (three of the twenty-six that he taught at the Ruskin), he brought in surgeons and philosophers, and other remote specialists to expose our young minds to astonishing things beyond the realm of art or art history. During one term we were taught about the shifting location of the soul inside the body and the ever-fraught historic tension between science and religion. In hindsight I found his way of teaching hugely impactful and inspiring. He informed my work and set me on the trajectory that I'm still very much on today. My connection to him feels very personal but, in reality, I was one of thousands of students he mentored and inspired.

Preparations for the *Cascading Principles* show and accompanying talks go back to 2018 and Brian's methods of constantly cross-fertilising, which were the templates for the talks and really for the show itself. Brian was meant to be part of these talks but tragically he died in 2022. However, it is his example that I tried to bring to this exhibition, a point at which art and mathematics can be brought into dialogue, provoking a cascading set of new principles and divergences. All too often, especially in the current political (and funding) environment, these disciplines can be pitted against one another and consequently siloed but their crossovers and cross-fertilisations are endlessly intriguing and productive. In Oxford itself, the Ruskin – an autonomous art school operating within the university, is in a unique position to allow its students to cast their nets wide while also being able to question, agitate, and disrupt.

I loved the chance to attend lectures in other subjects and wish I had attended more. It has certainly been intimidating to put my work in front of mathematicians and scientists, much less discuss these pieces with them, but the conversations and collaborations that have come out of this project have been so exciting; when you lose control of the meaning of something it is when you know that something truly poetic has happened. The exhibition began with initial talks about a commission back in 2014, which became a proposal to display a small series of my *Paradigms*, then ended up with the final exhibition of nearly 40 works covering almost all of my series to date. This has been in large part due to the influence and enthusiasm of Dyrol Lumbard and the joyful and profound curation of Fatoş Üstek, both of whom have gone way beyond the call of duty with this show; to both of them I am forever grateful. Credit here should also go to my trusty Ford Capri, with which I delivered a lot of the work in the exhibition. I bought the car whilst a student at the Ruskin and so I have owned it for nearly 25 years.

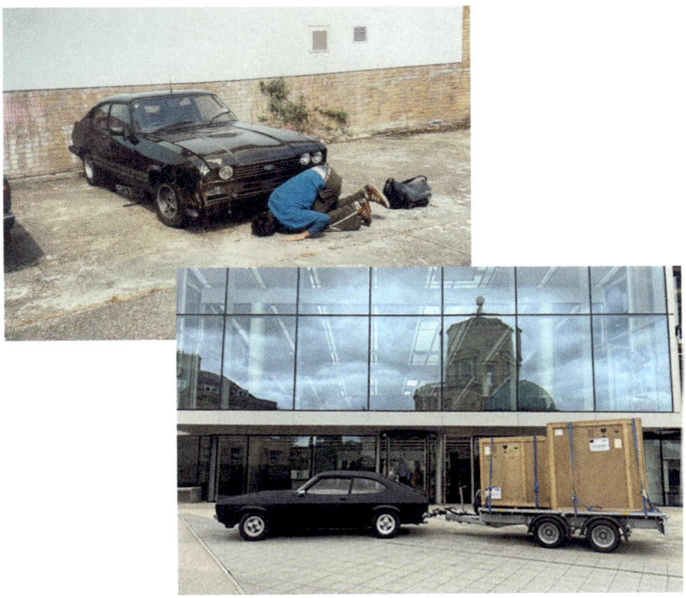

Capris past and present

In these lectures and the contributors' writing around them, I discovered new ways of seeing my own pieces and the ideas that they have generated. Mathematics and scientific ideas have always been at the heart of my practice, informing my basic unit of creation, the four-sided tetrahedron, the unit that grew into the stacked blocks of the *Paradigms*, then shattered into the leaves of *Fracture*, and expanded like fractal patterns into the branches of the *Formation Study*. These lectures allowed for the chance to discuss these pieces and the ideas surrounding them, as well as the crossover between art and mathematics (and indeed other disciplines). I was very honoured to take part in all of these talks but was particularly terrified to be in discussion with Sir Roger Penrose, and Carlo Rovelli. With each and every one of the contributors, a new conversation and new ideas were sparked. This book is the culmination of these explorations, and I hope the beginning for the reader's own.

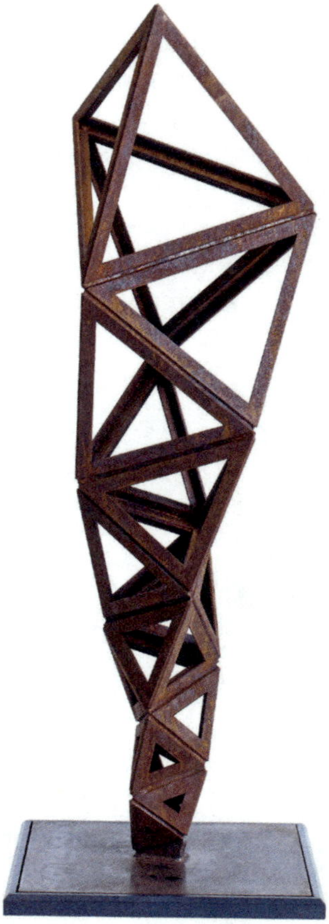

Paradigms

The writing of the scientific philosopher Thomas Kuhn inspired the title of the work, as he described in his book, 'The Structure of Scientific Revolutions', that "in order for ideas to progress old paradigms need to be toppled by new ones."

Built as a spiralling stack of tetrahedral steel blocks that increase in size by a set percentage at every twist, Shawcross explained that the *Paradigm* "is a beacon for progress and endeavour, but contains fallibility and should serve as a constant reminder of the precariousness of knowledge." These pieces grew from the tetrahedrons of the *Axiom* series and into the controlled explosions of the *Fractures*, and were born out of Shawcross' research into his commission for the Francis Crick institute, London in 2014.

All photographs are taken by Richard Ivey and show works by Conrad Shawcross unless otherwise stated.

Cover image
Slow Arc Inside a Cube XI, 2020, powder-coated steel, light, mechanical system

pp. 1, 23
Schism (Black Oak), 2022, wood, steel

p. 2
Lattice Cube 3, 2008, powder coated aluminium, nylon, stainless steel bolts

p. 3, left to right
Perimeter Studies (Icosahedron - full solid)
Perimeter Studies (Icosahedron - open face)
Perimeter Studies (Icosahedron - open side)
Perimeter Studies (Icosahedron - full structural), all 2021, all weathering steel

pp. 4, 5, 33, 76
Off Cut Axiom, 2009, douglas fir, steel

p. 6
Left cabinet, top to bottom:
Perimeter Studies (Octahedron - structural)
(Icosahedron - structural)
(Tetrahedron - structural)
(Dodecahedron - structural)
(Cube - structural)

Right cabinet, top to bottom:
Perimeter Studies (Octahedron - solid)
(Icosahedron - solid)
(Tetrahedron - solid)
(Dodecahedron - solid)
(Cube - solid)

All *Venice Presentation*, 2021
all weathering steel

p. 7
Schism (F45BB2), 2019, anodised aluminium Installation view at Victoria Miro, Frieze London, 3–6 October 2019, Courtesy the artist and Victoria Miro. Photo by Jack Hems.

p. 8
Bicameral Study, 2019, weathering steel

p. 13
Paradigm Chamfer (Structural), 2015
weathering steel

p. 22
Paradigm Chamfer (Solid), 2015
weathering steel

p. 24
Schism (Château La Coste), 2020
weathering steel. Photo by Rachel Taylor.

p. 26
Schism (Château La Coste), 2020
weathering steel. Photo by Daniel Kennedy.

p. 32
Space Tessellation (Triple Helix), 2005, oak, stainless steel. Photo by Conrad Shawcross.

p. 33
Axiom (Tower), 2009, douglas fir, steel
Installation view at Ministry of Justice.
Photo by Conrad Shawcross.

p. 35
Paradigm Broad (Structural), 2016
Paradigm Chamfer (Solid), 2015

Paradigm, 2016, all weathering steel
Photo by Marc Wilmot.

p. 37
Paradigm Broad (Solid), 2016
weathering steel

pp. 37, 110, 128
Fracture (R24H12), 2019
weathering steel, stainless steel stem

pp. 38, 129
Fracture (R45S33) - Enwrought Light, 2021
stainless steel

p. 39
Beacons (he15D6), *(om15D7)*, *(mo15D8)*, *(eh15D9)*, 2021
painted steel, oak, and aluminium

p. 40
The Dappled Light of the Sun, 2015, weathering steel. Installation view at Chatsworth House. Photography copyright Sotheby's.

The Dappled Light of the Sun, 2015
Photo by Marc Wilmot. Installation view at the Royal Academy of Arts, London.

p. 51
Images courtesy of Sir Roger Penrose. Initially used in 'Fashion, Faith, and Fantasy in the New Physics of the Universe' (Princeton University Press, 2016).

pp. 54, 151
Slow Arc Inside a Cube XI, 2020, powder-coated steel, light, mechanical system

p. 55
Image courtesy of Sir Roger Penrose. Initially used in 'The Road to Reality: A Complete Guide to the Laws of the Universe' (Jonathan Cape, 2004).

p. 57
Graph by Carlo Rovelli. Initially used in 'Fashion, Faith, and Fantasy in the New Physics of the Universe' (Princeton University Press, 2016).

p. 60
Image by Francesca Vidotto. From 'Reality Is Not What It Seems: The Journey To Quantum Gravity', Carlo Rovelli (Allen Lane, 2016).

p. 64, left to right
M.C. Escher's "Circle Limit I" copyright 2024 The M.C. Escher Company, The Netherlands; M.C. Escher's "Circle Limit I with lines" copyright 2024 The M.C. Escher Company, The Netherlands. All rights reserved. www.mcescher.com

pp. 68–70
Reprinted article: 'Gravitational Collapse and Space-Time Singularities' with permission from American Physical Society as follows: Sir Roger Penrose, Physical Review Letters, vol. 14, Issue 3, 18 January 1965, pp. 57–9. Copyright 1965 by the American Physical Society.

p. 77
Liliane Lijn, *Zero Gravity Koan*, 2004
glass reinforced polyester, Perspex, motorised drive and fluorescent lights, 250 × 150 cm base diameter. Niccone Valley, Italy. Photo by Liliane Lijn. Courtesy the artist and Rodeo, London/Piraeus.

p. 78
Liliane Lijn, *Split Spiral Spin*, 1982
perforated stainless steel and zinc coated mild steel, 1159 × 610 cm base diameter. Commissioned public sculpture for the Birchwood Science Park. Photo by Stephen Weiss. Courtesy the artist and Rodeo, London/Piraeus.

p. 79
Liliane Lijn, *Gravity's Dance*, 2019
black weighted cloth, LEDs, Forex central stiffener, 3 phase motor, inverter, control system, 600 cm diameter. Photo by Stephen Weiss. Installation view, 'Il Ritmo dello Spazio' / 'The Rhythms of Space' at Museo della Grafica di Pisa. Courtesy the artist and Rodeo, London/Piraeus.

pp. 80, 81
Liliane Lijn, *Temenos*, 2021
painted steel poles, painted galvanised base ring, LEDs, 1130 × 780 cm. Commissioned public sculpture for King's Cross Central Limited Partnership. Photo by John Sturrock. Courtesy the artist, King's Cross Central Limited Partnership and Rodeo, London/Piraeus.

p. 82
Liliane Lijn, *Liquid Koan*, 2003
translucent resin, 30 × 12 cm base diameter. Photo by Liliane Lijn. Courtesy the artist and Rodeo, London/Piraeus.

p. 83, top to bottom
Liliane Lijn, *Flow Lines*, 1975, 21 × 26 cm, ink on paper. Photo by Stephen Weiss. Courtesy the artist and Rodeo, London/Piraeus; Liliane Lijn, *Carbon Black*, 1988
457 × 610 × 122 cm, painted perforated mild steel. Commissioned public sculpture for The Government Chemist. Photo by Stephen Weiss. Courtesy the artist and Rodeo, London/Piraeus.

p. 84
Loop System Quintet, 2005
Courtesy the artist, The New Art Gallery, Walsall, and Victoria Miro, London. Photo by Conrad Shawcross and Jonathan Shaw.

p. 85
Manifold (Major Third) 5:4, 2023
Liverpool Street station (Elizabeth line). Commissioned as part of The Crossrail Art Programme. Copyright Conrad Shawcross. Courtesy the artist and Victoria Miro. Photo by GG Archard.

p. 88
Slides courtesy Marcus du Sautoy.

p. 90
Schism (F24YB3), 2020
anodised aluminium

pp. 94–5
Images courtesy Fatoş Üstek.

p. 97
Patterns of Absence "M" (Bm15D7), 2021
painted steel, mechanical system

p. 98
Library of Absence (Umbrette), 2023
aluminium, stainless steel, nylon, custom electronics, paper

p. 104
Liliane Lijn, *Two Worlds*, 1959
31.7 × 24 cm, ink on paper. Photo by Stephen Weiss. Courtesy the artist and Rodeo, London/Piraeus.

p. 104
Liliane Lijn, *Split Spiral Spin*, 1982
1159 × 610 cm base diameter, perforated stainless steel and zinc coated mild steel. Commissioned public sculpture for the Birchwood Science Park. Photo by Stephen Weiss. Courtesy the artist and Rodeo, London/Piraeus.

p. 112
A phase portrait of the Riemann zeta function. Copyright Professor Elias Wegert. Reproduced with his kind permission. For further details see 'Visual Complex Functions: An Introduction with Phase Portraits', Elias Wegert (Birkhäuser, 2012).

p. 114
Covers of volumes 1–8 of *Collapse*, 2007–2014, Urbanomic Media Ltd.

p. 118
Ruth Asawa: Citizen of the Universe, installation view at Modern Art Oxford, 2022. Photo by Ben Westoby. Artwork copyright 2024 Ruth Asawa Lanier, Inc. / ARS, NY and DACS, London. Courtesy David Zwirner.

p. 120
Collage of Ruth Asawa Portraits at Black Mountain College, North Carolina, c. 1948–1949. Photo copyright Estate of Elizabeth Jennerjahn. Courtesy Ruth Asawa Lanier, Inc.

Buckminster Fuller and students, Summer Institute, Black Mountain College, 1949. North Carolina Museum of Art Black Mountain College Research Project. Courtesy of the Western Regional Archives, State Archives of North Carolina.

p. 121
Ruth Asawa, *Untitled* (BMC.144, Dancers), c. 1948–1949, blue coloured pencil and watercolour on tracing paper, 8 × 10½ in. (20.3 × 26.7 cm). Achenbach Foundation for Graphic Arts. Artwork copyright 2024 Ruth Asawa Lanier, Inc. / Artists Rights Society (ARS), New York. Courtesy David Zwirner.

p. 122, top to bottom
Ruth Asawa holding up a milk carton construction of a globe at McKinley Middle School in Redwood City, California, 1972. Courtesy Ruth Asawa Lanier, Inc.; A student using milk cartons to make a sculpture, 1981. Photo by Anne B. Page, Courtesy Ruth Asawa Lanier, Inc. / Fine Arts Museums of San Francisco.

p. 128
Bicameral Study, 2020, aluminium

p. 131
Paradigm Slender (Structural), 2016 weathering steel

p. 138, 159
Bicameral Study, 2019, anodised aluminium

pp. 153
Paradigm Broad (Structural), 2016 weathering steel; *Paradigm Chamfer (Solid)*, 2016 weathering steel

p. 154, left to right
Paradigm Slender (Structural)
Paradigm Chamfer (Structural)
both 2015, weathering steel

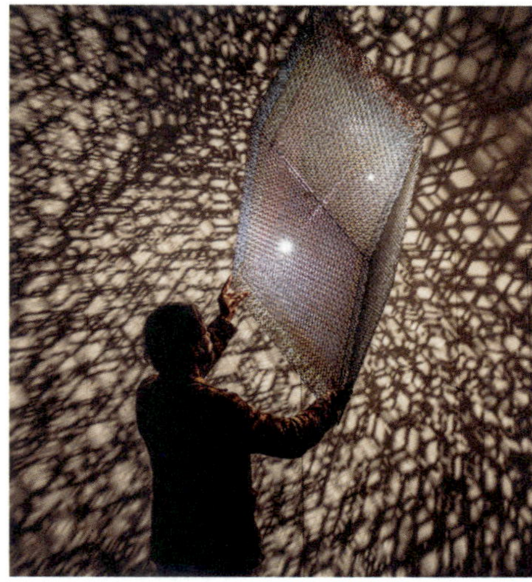

p. 155
Paradigm (Solid), 2015, weathering steel

p. 156
Paradigm Study (Solid), 2014 weathering steel

p. 157
Paradigm Slender (Solid), 2015 weathering steel

p. 158, top to bottom
Beacons (he15D6), (om15D7), (mo15D8), (eh15D9), 2021, painted steel, oak, and aluminium; *Formation Study IV (Prevailing Wind)*, 2018, weathering steel

p. 160
Plosion 1 (Yellow) Large, 2013 aluminium, glass, wood

An Archivorum×M artist's book, a publishing series conceived for Magic Megève. Published on the occasion of Conrad Shawcross' major UK survey exhibition, curated by Fatoş Üstek at the Mathematical Institute, Oxford University, which opened in September 2022.

Assistant Editors:
India Lewis, Phoebe Howard

Design:
Fraser Muggeridge studio

Published and distributed by:
Mousse Publishing
Contrappunto s.r.l.
Via Pier Candido Decembrio 28
20137, Milan–Italy

Available through:
Mousse Publishing, Milan: moussemagazine.it
Archivorum: archivorum.org

First edition, 2024
Printed in Belgium by Graphius
ISBN 978-88-6749-635-8
€35 / $40

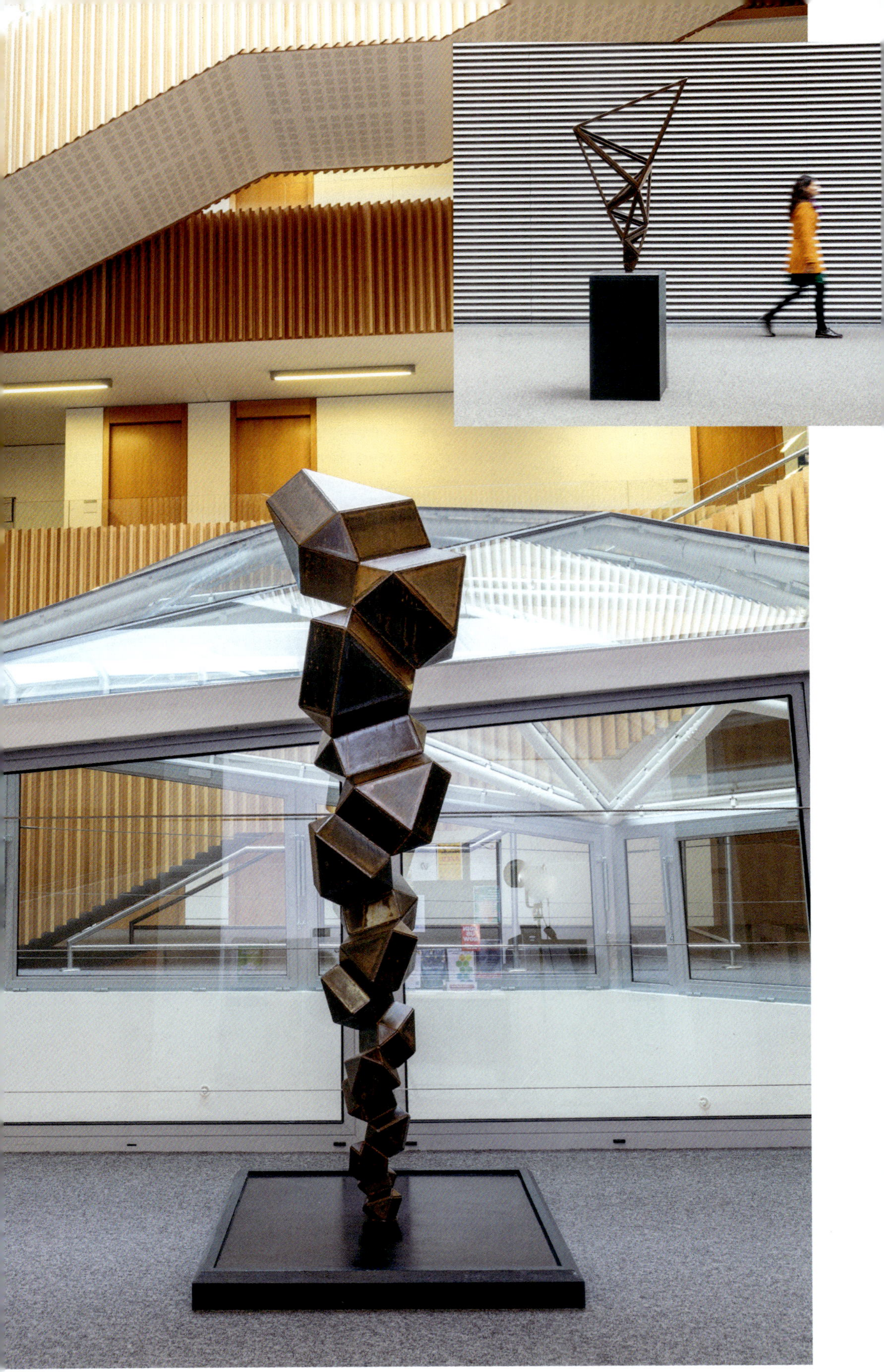

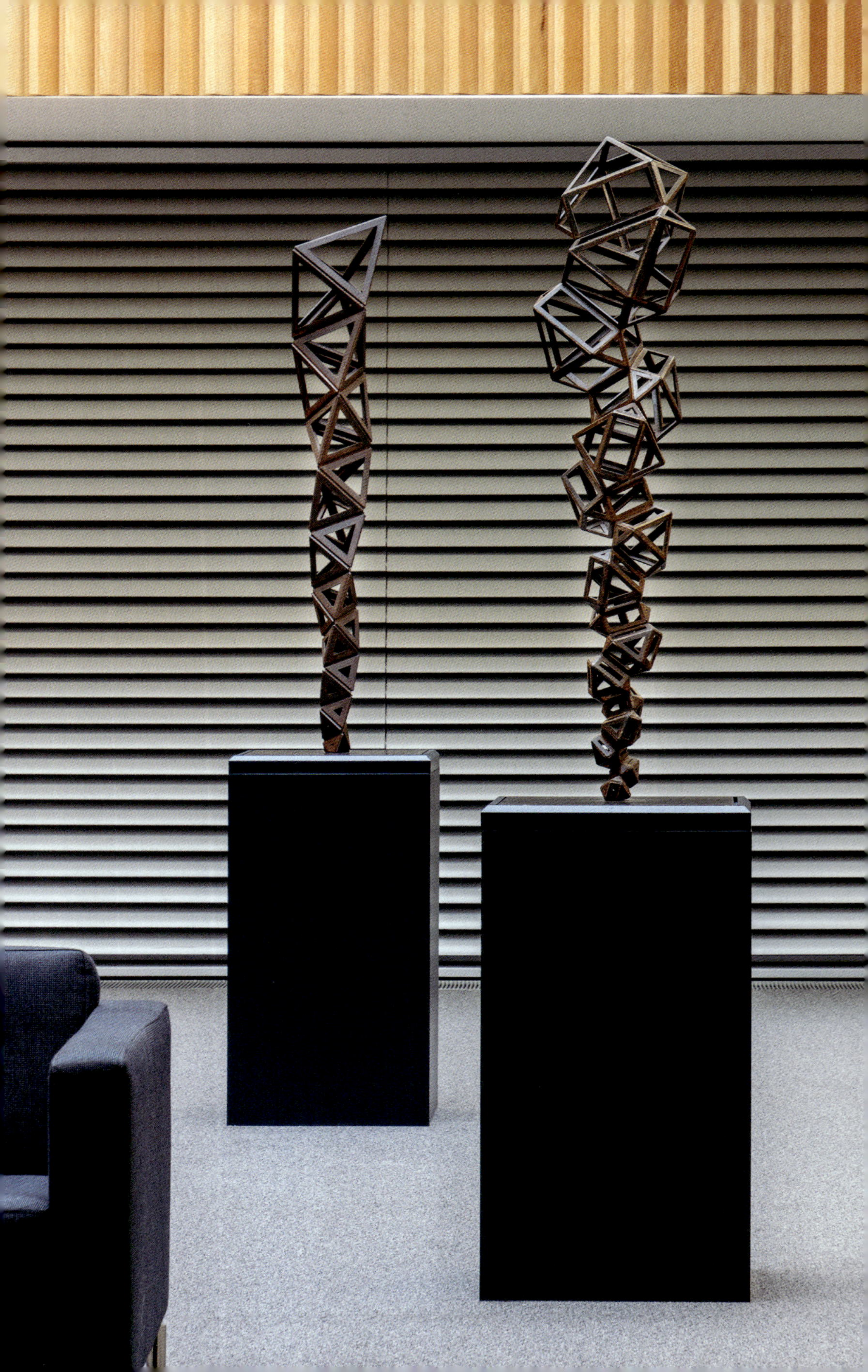

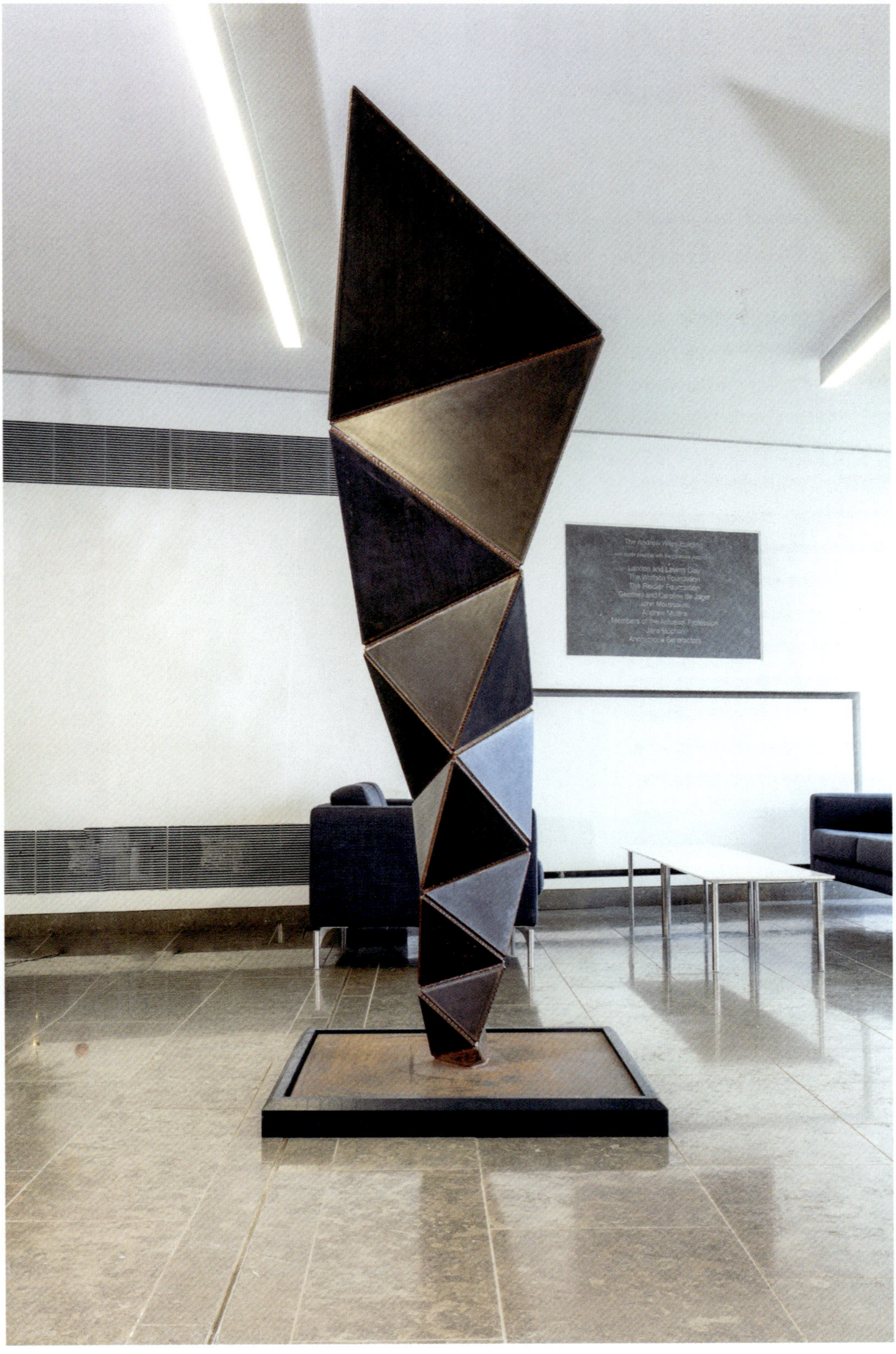

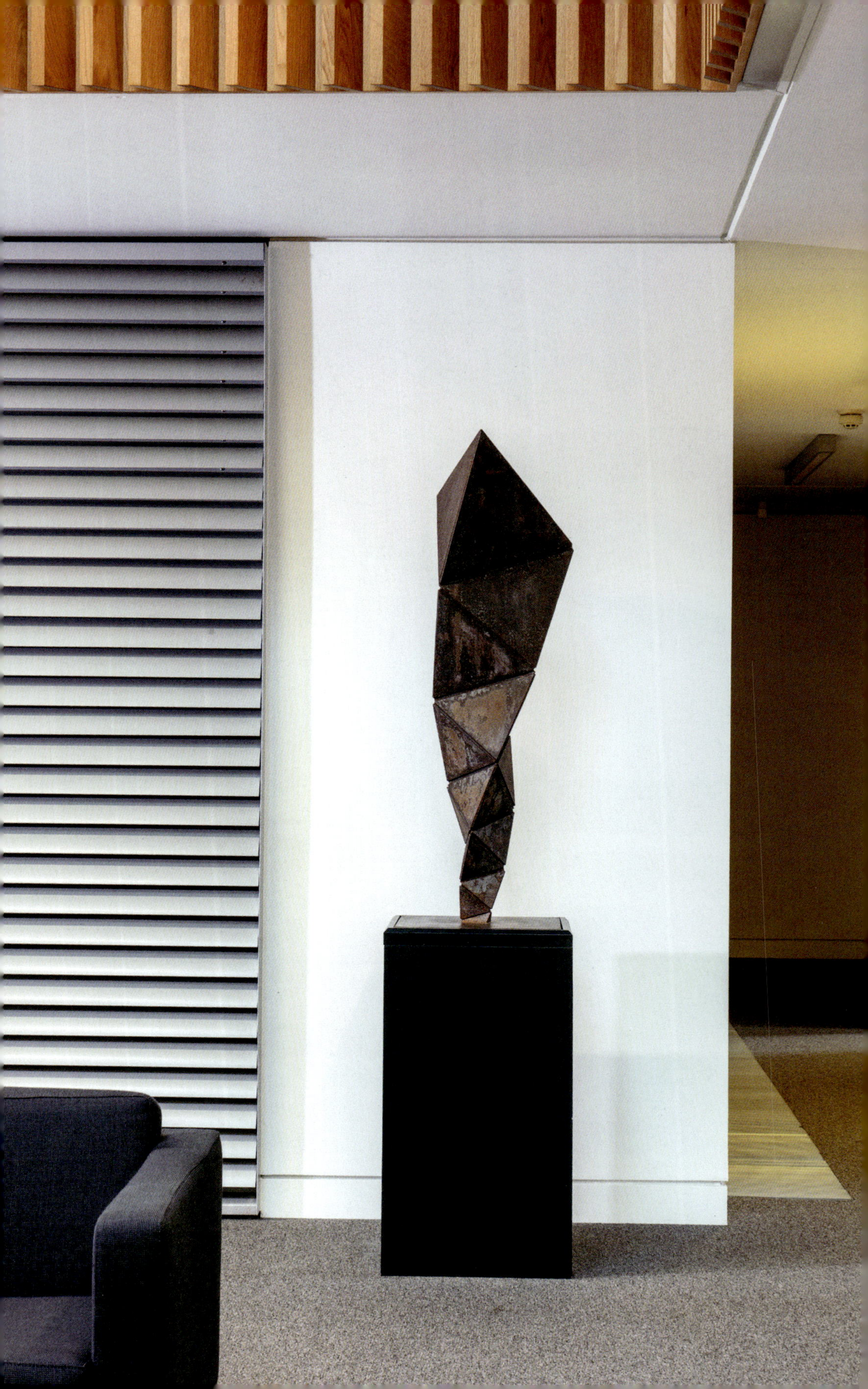

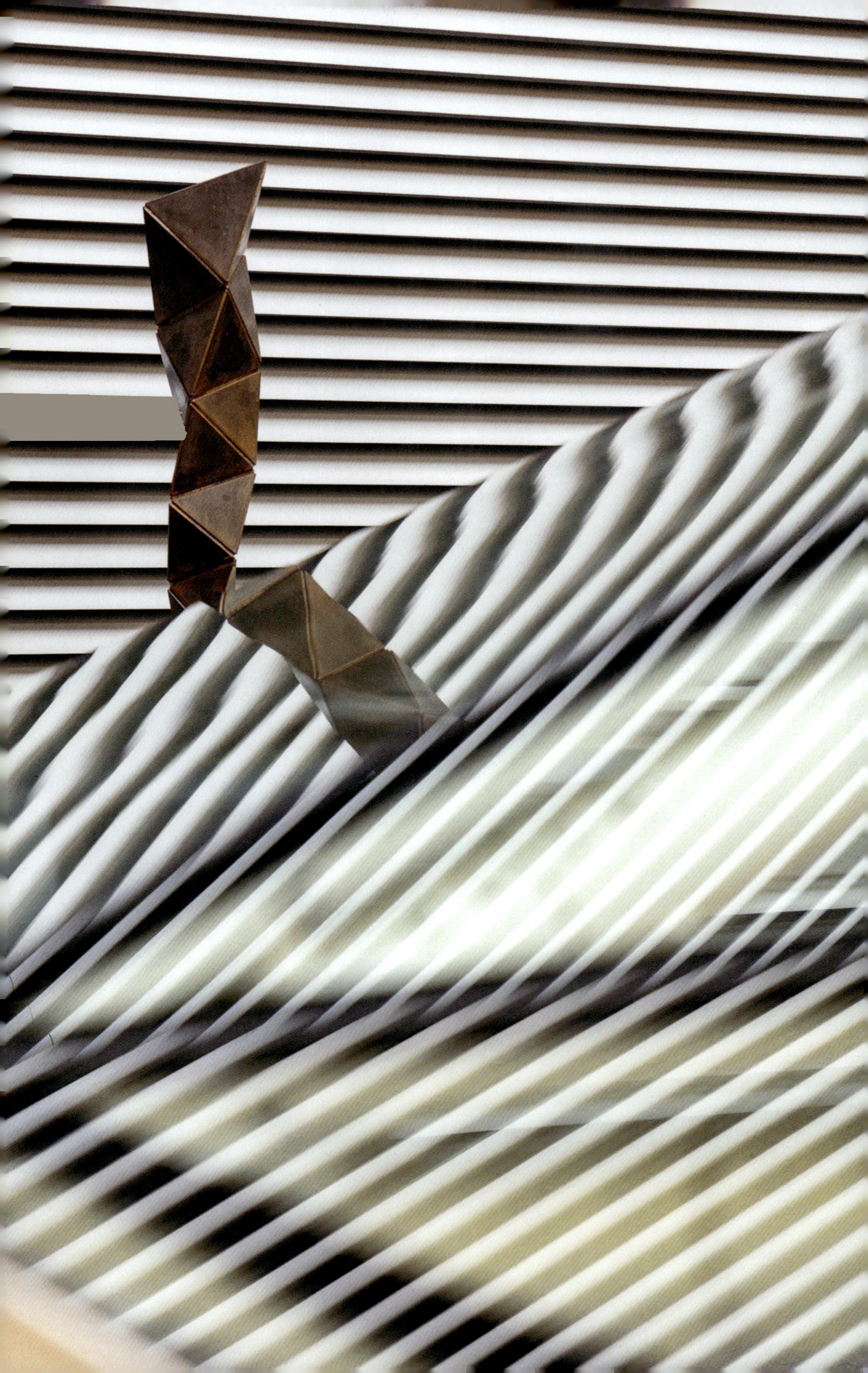

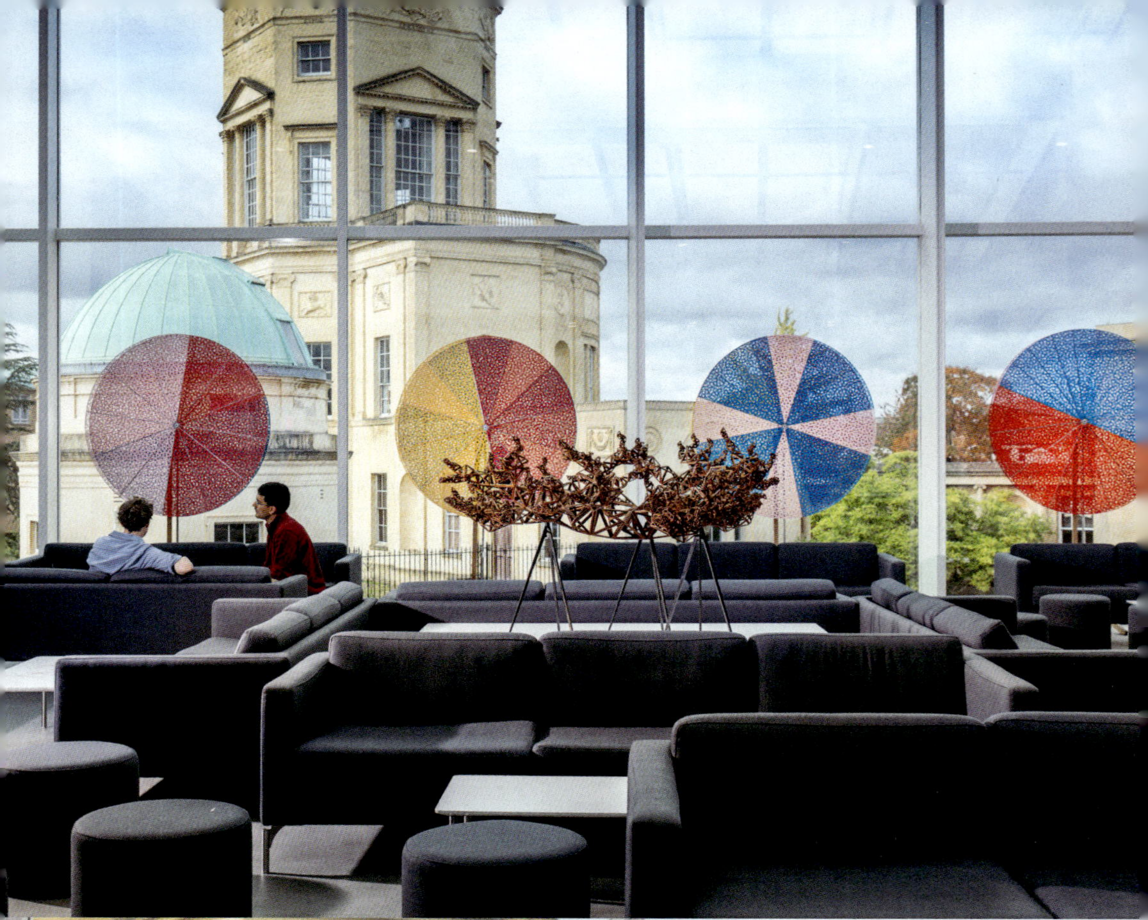

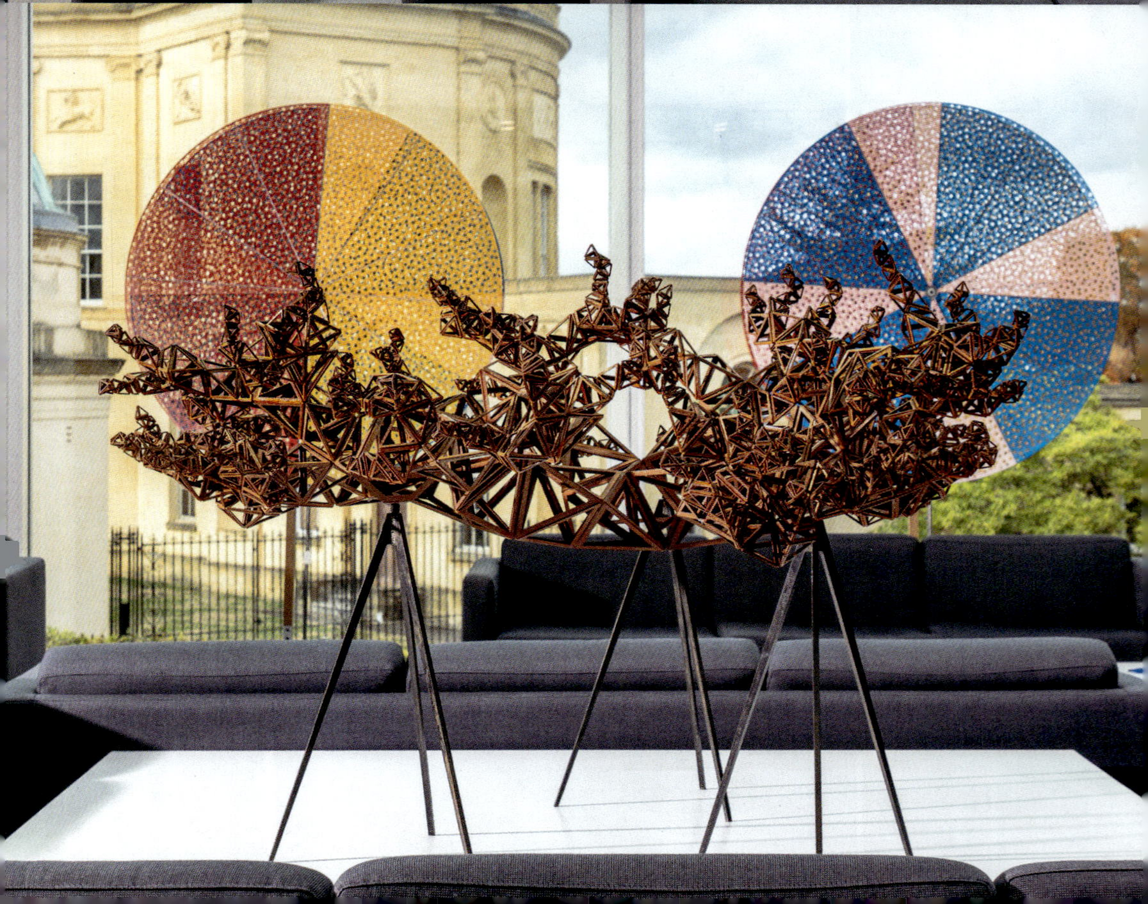

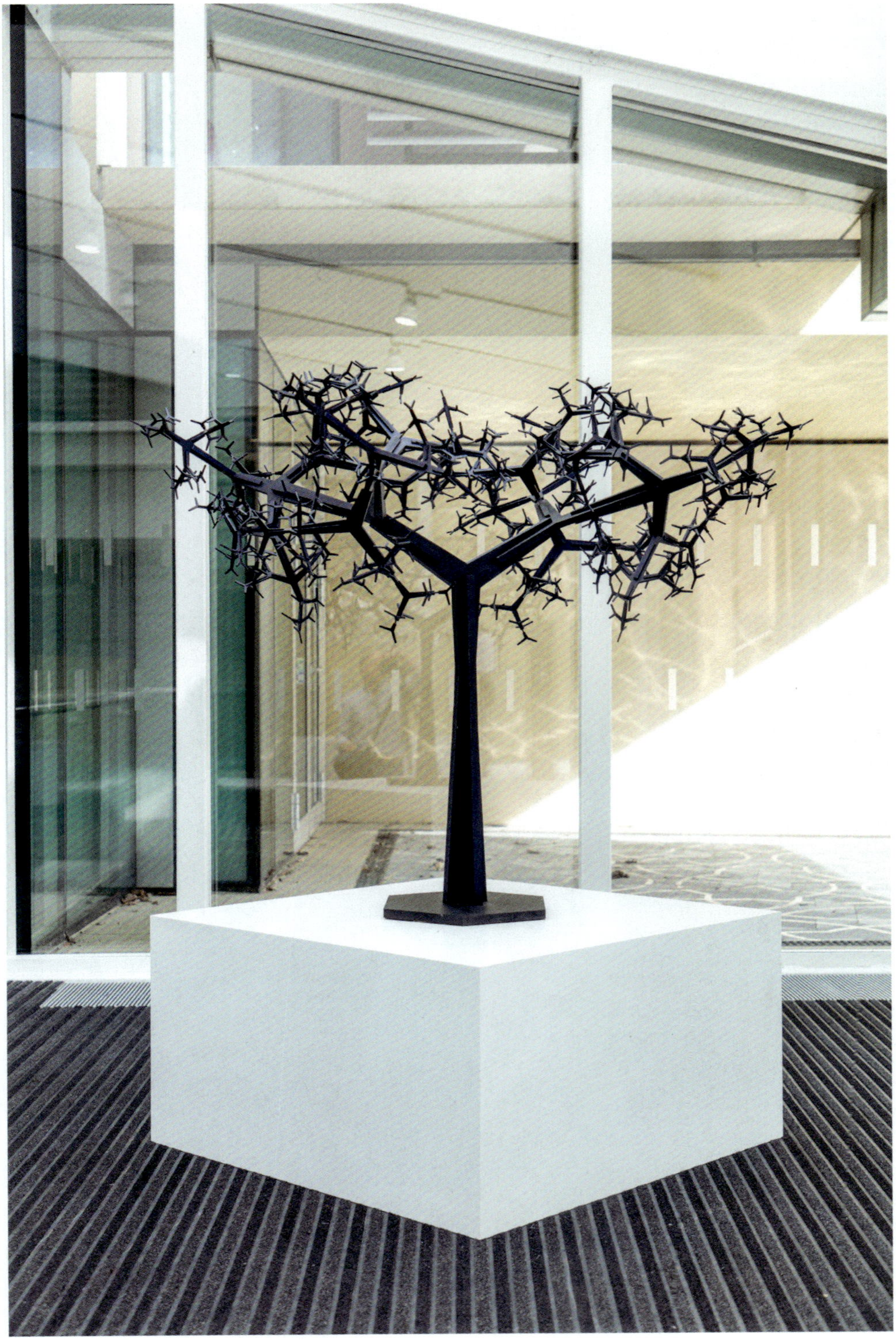

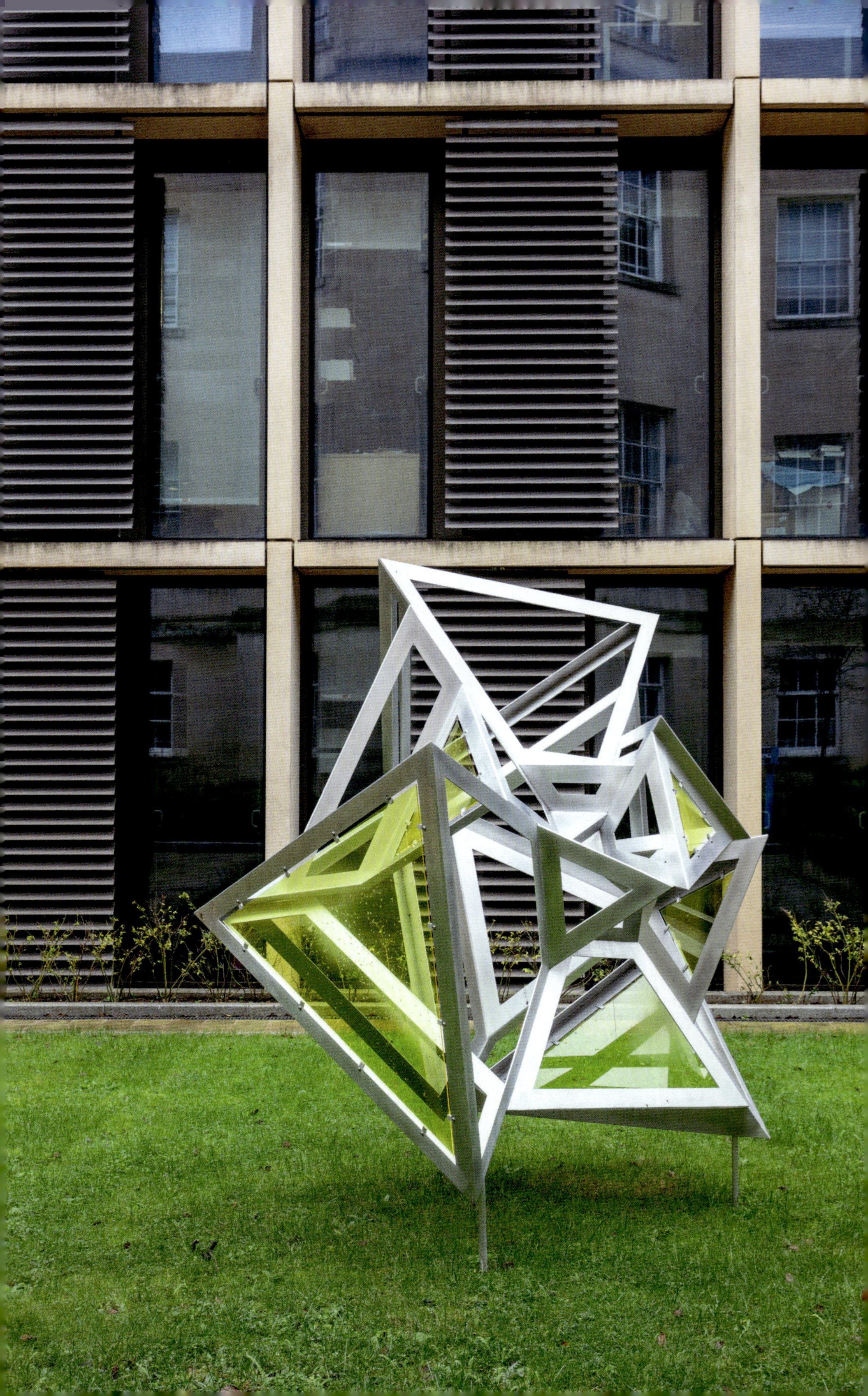